SOGGY GROUND

SOGGY GROUND
A GEOGRAPHY OF PINE BARRENS WETLANDS

By

Mark Demitroff

South Jersey Culture & History Center

2024

Soggy Ground: A Geography of Pine Barrens Wetlands.

First Edition published in 2024.

Copyright © 2024 by Mark Demitroff.
Copyright © 2024 photographs owned by other than the author.
Design and layout Copyright © 2024 by the South Jersey Culture & History Center at Stockton University.

All rights reserved. No part of this work may be reproduced or transmitted in any form by any means, electronic or mechanical, including photocopying and recording, or by any information storage or retrieval system without permission in writing from the publisher, except in the case of brief quotations embodied in reviews and certain other non-commercial uses permitted by copyright law.

ISBN: 978-1-947889-21-7

Jacket Cover. An excerpt of The Lake Tract survey (Clark & Hitchnor 1802) in geographic reflection marries the old with the new by superposition upon Great Egg Harbor River watershed laser altimetry (BoydsMaps, Terrain Viewer, ©2023 Boyd Ostroff, reprinted with permission). Designed by Gary Schenck.

stockton.edu/sjchc/

Acknowledgments

My quest for evidence of land change is rooted in the experiential knowledge generously shared by many Pine Barrens' citizens. Local farmers, hunters, botanists, historians, gravel pit operators, and other individuals too numerous to name have collectively chronicled the spirit of this special place. Their insights into the complex physical and cultural dynamics of Pinelands National Reserve are important compliments to scientific knowledge and can provide key historical information. Local, long-term expertise should not be summarily dismissed.

But traditional knowledge only takes one so far; science is based on empirical evidence. This story would have been impossible without Hugh French's knowledge of the periglacial processes, past and present. I have received enlightenment from my former friend, colleague, and mentor.

The Great Egg Harbor Watershed Association (GEHWA), through the National Park Service's Wild and Scenic Rivers Program, provided critical seed grants to foster this research; with notable credit extended to GEHWA President Julie Akers, Administrator Fred Akers, and the trustees. My fellows at the West Jersey Historical Roundtable provided essential archival and moral support, with special thanks to Claude Epstein, Paul W. Schopp, Ed Fox, Robert Barnett, Bill Leap, Bill Farr, Joan Soderlund, Bonny Beth Elwell, and Gabe Coia.

Wayne Newell, Fritz Nelson, Delphis Levia, Michael Hozik, and Mark Mihalasky contributed early encouragement and—along with later cooperation—Stephen Wolfe, Dorothy Merritts, Robert Walter, Ben DeJong, Dmitry Streletskiy, Victor Konishchev, Victor Rogov, George Gao, Malcolm LeCompte, Matti Seppälä, Barbara Woronko, Dorota Chmielowska, Ilya Buynevich, Charles Tarnocai, Hans-Peter Blüme, Mike Walagur, Jim

Doolittle, Julie Laity, Ron Dorn, Scott Stanford, Russell Losco, Tom Meierding, Scott Andres, and Scott Keenan—all who shared their geotechnical expertise. Steven Forman, University of Illinois at Chicago, generously provided the initial optical dating and insight on eolian processes along with Tammy Rittenour, Sebastien Huot, Shannon Mahon, and Olav Lian.

Jack Cresson spent many hours guiding me to various precontact and other interesting archeological sites across South Jersey, with further assistance from Joseph Arsenault, Alan Mounier, Budd Wilson, Ted Gordon, Carl Farrell, Richard Regensburg, Emile DeVito, Ronny Bertanazzi, Mike Cicali, Jamie Cromartie, Joseph Smith, Patt Martinelli, Kent Mountford, Pierre Lacombe, Robert Kecskes, George Zimmerman, Matt Tomaso, Rich Bizub, Renée Brecht-Mangiafico, Guy Thompson, Bob Moyer, Terry Schmidt, Willy Hamilton, James Pullaro, Robert Wells, Wayne Russell, Seymour Jones, Jr., and Mary-Ann Thompson; all who aided me with botanical and/or geographic aspects of this document. Harry and Gail Benson aided with early field, computer, and historical assistance.

Abriana Bradley and Tom Kinsella are thanked for their invaluable editorship, which has made for a much more effective story to emerge. Their task was not an easy one given the obscurity and the technical nature of the material presented within. Kudos!

Last, but not least, I extend my sincerest gratitude to my wife Patricia, who tirelessly spent countless hours on graphics preparation, fieldwork coordination, and research mobilization—in company with my treasured daughter Alexis. I thank them for both their time and patience through this decades-long process. I owe you.

Contents

ACKNOWLEDGMENTS 5
Chapter 1. INTRODUCTION 9
    Pine Barrens Overview 9
    Physical, Cultural and Historical Dynamics 11
    The Woodsman of the Pines 14
    Edaphics: How the Sandy Soils have Influenced Living Things . 15
    Shallow Groundwater 16
Chapter 2. PINE BARRENS GEOLOGY AND GEOMORPHOLOGY 21
    Geologic Framework 21
    Contested Pleistocene Inheritance 21
    The Frozen Landscape of Newell and Others 26
    Palynological Evidence 27
    Renewed Interest in the Periglacial Environment 28
    Epilog to the Spirited Debate 30
Chapter 3. HOLOCENE HYDROGEOGRAPHIC FEATURES AS
    A RESOURCE FOR CULTURAL ECOLOGY 53
Chapter 4. SPUNGS OF THE PINE BARRENS 59
    Etymology of Spung 59
    Appellatives of Spungs 61
    Spung Morphology 63
    Early Spung Use and Topography 65
    Origin of Spungs 68
    Spungs: Summary 73
Chapter 5. CRIPPLES OF THE PINE BARRENS 101
    Etymology of Cripples 101
    Appellatives of Cripples 103
    Cripple Morphology 103
    Early Cripple Use and Topography 105
    Origin of Cripples 105
    Cripples: Summary 108

Chapter 6. BLUE HOLES OF THE PINE BARRENS ... 125
- Etymology of Blue Holes ... 125
- Stories of Blue Holes ... 126
- Blue Hole Morphology ... 128
- Early Blue Hole Use and Topography ... 129
- Origin of Blue Holes ... 131
- Blue Holes: Summary ... 137

Chapter 7. SAVANNAHS OF THE PINE BARRENS ... 157
- Etymology of Savannahs ... 157
- Appellatives of Savannahs ... 160
- Morphology of Savannahs ... 161
- Early Use of Savannahs and Topography ... 162
- Of Savannahs and Plains ... 164
- Origin of Savannahs ... 166
- Savannahs: Summary ... 170

Chapter 8. WINDOWS TO THE GROUND WATER ... 191
- A Dwindling Natural Resource ... 191
- Spungs as Indicators of Hydrologic Change ... 193
- Broad Pond ... 194
- Egg Harbor Pond ... 195
- Spung at Campbella ... 196
- Horse Break Pond ... 196
- Lochs-of-the-Swamp ... 198
- Cripples as Indicators of Hydrologic Change ... 199
- Springs as Indicators of Hydrologic Change ... 200
- Savannahs as Indicators of Hydrologic Change ... 200

Chapter 9. SUMMARY ON SPUNGS, CRIPPLES, BLUE HOLES, AND SAVANNAHS ... 215
- Ice Age Inheritance ... 215
- Water Resources ... 217
- Preservation of Our Periglacial Heritage ... 220
- Conclusion ... 222

REFERENCES ... 225

01
Introduction

Pine Barrens Overview

In the midst of megalopolis lies a half-million-hectare (1.2 million acre) tract known as the Pine Barrens (Fig. 1.1). Thick forests of pitch pine (*Pinus rigida* [P. Mill.]) and mixed oak (*Quercus* spp.) sweep across the broad, flat uplands of New Jersey's Outer Coastal Plain. Lowland swamps of Atlantic whitecedar (*Chamaecyparis thyoides* [L.]) and mixed hardwood dissect the sandy and gravely sediments of the uplands, creating a labyrinth of lush wetlands that drain the parched barrens positioned just above them. Between 25% (Ehrenfeld 1986) to 35% (Roman & Good 1983; Zampella & Roman 1983) of the Pinelands[1] are classified as wetlands. It is truly remarkable that such sparsely populated woodland, described as "a botanist's wonderland, a developer's dream, a naturalist's last stand" (Pearson 1994: 141), could remain relatively intact, encircled by suburban sprawl.

Preserved within the Pinelands' bounds are four distinct hydro-geographic features known colloquially as *spungs*, *cripples*, *blue holes*,

---

1. Following Boyd (1991), the terms Pine Barrens, The Pines, and Pinelands are used interchangeably for sake of variety. This exchange is common in everyday language. In the original ecological sense (Harshberger 1916: 8–9; Wacker 1979: 3–9), *Pine Barrens*, *Pine Lands* and *The Pines* described a poorly defined sandy region of New Jersey's Outer Coastal Plain. In 1978 and 1979, state and federal authorities adopted *Pinelands* to specifically designate an area conserved as a National Reserve. Not all portions of the Pine Barrens are protected under this jurisdiction, and the new legislative boundary extends to limited parcels just outside the Pine Barrens. Due to its sandy nature, pine (*Pinus* spp.) codominates and the land is barren for agricultural purposes—hence Pine Barrens.

and *savannahs*, which form the basis for this book. Little is known of the nature or extent of these hydrogeomorphic wetland forms. To early woodsman, they denoted important hunting grounds, watering spots, and places where abundant forest resources could be foraged. The original meanings of these anecdotal expressions are long forgotten, as many of the features have been modified or degraded beyond recognition. Development, and an apparent lowering of the regional water table, imperils remaining features (French & Demitroff 2001; NJDEP 2003a, 2003b, Bizub 2014a, 2014b; Kecskes 2014).

Within the wetlands of the New Jersey Pine Barrens are many of the region's rare, threatened, and endangered species (Cromartie 1982; Beans & Niles 2003; Juelg 2003, 2014; CFWNJ 2021). The unique flora and fauna have been scrutinized and lauded by the scientific community for over a century (Harshberger 1916; Small 1952; Forman 1979a; Snyder & Vivian 1981, Boyd 2008; Bunnell et al. 2018). Closed basins (*spungs*) support rare insect populations, including the chain fern borer moth (*Papaipema stenocelis*) and the chalky wave (*Scopula purata*) (Dowhan et al. 1997). The seriously threatened bog asphodel (*Narthecium americanum* [Ker Gawler]) occurs exclusively in the diminishing savannah habitat of the New Jersey Pine Barrens (Schuyler 1990, Kelly 2018).

In the last forty years, archeologists have begun to recognize the importance of these wetlands (Bonfiglio & Cresson 1982; Mounier 2003; Cresson et al. 2006; Tomaso et al. 2008; Lattanzi 2017) for the artifacts found in association with them after 14,000 years (14 ka) of human visitation. Until recently, the Late Pleistocene climatic forces that shaped the Pinelands' land surfaces remained enigmatic. Taking into consideration recent geomorphological studies (French & Demitroff 2001; 2003; 2012; et al. 2003; 2005; 2007; 2009a; Lemcke & Nelson 2004; Newell 2005; Newell & DeJong 2011; Newell & Wyckoff 1992; et al. 2000), a long overdue account of the Pinelands' distinctive wetlands can now be advanced, more than twenty years after I began this book under the encouragement of soil scientist John Tedrow.

Physical, Cultural, and Historical Dynamics
About this Compendium

A common criticism of any historical science is that historians have a very difficult time proving anything. And if history cannot prove something then it must not be science (Martin 1998: 1).

Herein lies a dilemma. This geographer wishes to systematically document Pinelands landforms and to include local knowledge about a place with which he has a lifetime of intimate acquaintance. Experiential knowledge can be a useful complement to scientific knowledge (Fazey et al. 2013). This is an adventuresome pursuit, being the first paper to broadly link a paleo-periglacial landscape with both the cultural and environmental dynamics of this region. The current inquiry is not concerned with substantiating a hypothesis quantitatively. Geography's "quantitative revolution" of the 1950s and 1960s (Strahler 1950; 1952; Burton 1963; Harvey 1969; Cresswell 2014; Wolfe, L. J. et al. 2020) provided too narrow a focus for many subsequent geographers. Instead, its purpose is to show how human occupation made use of a unique landscape, and how later broad-scale human processes have obscured signs of these relationships. With a framework of previously unavailable climatic and surficial-process information, insight is provided on land change over a timescale outside the purview of most recent research endeavors.

Admittedly, this work strays from the rigors of "realist" scientific theory of Plank and Kuhn (Fuller 2005). The current foundational, authoritarian, and risk-averse philosophy of positivism does not interface well with collecting data from the communal wisdom about historical and ecological aspects of local wetlands (Bassett & Zimmerer 2003; Myers et al. 2003; Perennou et al. 2018). Positivism left little room for bold conjecture about the nature of landforms, although 14,000 years of qualitative inference indicates that something might be there! Geomorphology is not restricted to applications of computer or laboratory modeling and remote sensing (Martin & James 1993: 412–428). The topic involves old-fashioned fieldwork and historical archival work (Barrows 1923; DeLyser & Starrs 2001; de Blij 2005: 49; Meyer et al. 2014; Meyer & Youngs 2019); let us not forget our geographic roots! A multiplicity of approaches and traditions are valued in geography because "there is no right way to comprehend humans on the land as our central mission" (Martin & James 1993: 457).

Introducing the concept of human ecology, Barrows (1923: 3) emphasized the importance of geographical relationships "between natural environments and the distribution and activities of man." He pointed out how humanizing physical geography is a complex task, requiring an understanding of multiple human–environmental relationships. If we accept that geography is the mother of sciences, including botany, zoology, geology, archeology, anthropology (Barrows 1923), then it is no surprise that geographers share interests with neighboring disciplines. As geographers, "we thrive on cross-fertilization and diversity" (Sauer 1956: 293), and to do so should in no way threaten the independence of kindred disciplines.

Sauer (1941) suggested that the study of long-term environmental adaptations by the traditional (i.e., folk) world provided better insight into the natural landscape than modern world interactions. He marveled at the geographic skills of native guides in Mexico, and their ability to interpret the "lay of the land" (Sauer 1956: 289–290). Later in life, Sauer wrote of his "warm affection for those folk who live contentedly and with enlightenment in a harmonious balance of people and place, of land and life" (Hart 1964: 614). According to Hart, it was through human geography (especially rural) that Sauer repeatedly addressed modern man's destruction of such natural resources as soil and water. The new discipline of *cultural ecology* evolved to address the complex "cultural, historical, political, and biophysical dimensions of human-environmental interactions" (Bassett & Zimmerer 2003: 107). Experiential knowledge is still considered a valuable complement to scientific knowledge (Bassett & Zimmerer 2003; Fazey et al. 2013). It is in this old-fashioned spirit that a truly important ecosystem can achieve value and appreciation in a modern context and be saved from the ravages of uncontrolled sprawl. The challenges of land-use change and sustainability science (Kates 2011) are nothing new; only the array of methodologies available to explain land use and its impacts has greatly expanded.

This study makes use of three themes to integrate several broad and seemingly disparate threads of geographical significance: 1) surficial deposits in the Pine Barrens have distinctive characteristics inherited during cold, dry, and windy episodes of the Late Pleistocene (126–11.7 ka); 2) the Ice Age landscape influenced Pinelands historical and cultural geography dynamics; and 3) understanding of these topics is relevant to modern issues of land preservation and use. The primary difficulty faced in such an understanding lies in bridging the very real gap between the physical and cultural aspects of geographical study.

To Thornthwaite (1940: 347), "even the most aggressive and ambitious of the geographers" wouldn't stray into human ecology and other social sciences. Rhoads (2004) states—in reference to geography—that the "mutual accommodation of human and physical sides of the discipline has proven difficult." This author concurs, and further agrees with Rhoads' assertion that physical geographers should collaborate more with other disciplines, and that to do so offers tremendous potential for recognizing and solving environmental problems (also see Sauer 1941; Sui & DeLyser 2012; DeLyser & Sui 2014). We can connect with our past by studying the spatial dimensions of natural and social processes, helping to define the identity of a place. A multidisciplinary geographical approach will assess ecosystem health and land changes and provide innovative decision-making tools to resource managers. Geography accommodates cross-disciplinary studies to integrate places with spatial interconnections (McMahon et al. 2005).

Recognition and understanding of the Pine Barrens hydrogeographical landforms (spungs, cripples, blue holes, savannahs) will raise questions about broader issues concerning the region and can add insight across a range of disciplines. To the geomorphologist, they are distinctive features that can be used as evidence of Ice Age frost action and frozen ground. To the biologist, the legacy of periglacial features provides critical habitat for many of the rare, threatened, and endangered plants and animals. To the climatologist, periglacial phenomena can be used to reconstruct Late Glacial to Holocene conditions with respect to temperature, precipitation, and wind patterns. To the archeologist, relict landforms are noteworthy as sites where artifacts of the Pinelands' earliest cultures can be found. To the historian, relict landforms can mark places of colonial settlement, the locus of early roads, and timberland and frontier commerce. To the land-use planner, recognition of periglacial landforms provides geotopes and biotopes worthy of preservation and information useful in efforts to control sprawl and create healthy, livable communities. To the hydrologist, they are windows into the shallow water table, and act as barometers of changing groundwater levels.

This work attempts to integrate material from disparate fields to help address questions about the sustainability of Pinelands environmental systems. Healthy skepticism exists about science's singular ability to transcend political and economic pressures (Haas 2004; van Kerkhoff & Lebel 2006; Grundmann 2017). Too often, key land-change decisions are made behind closed doors, with little public input. Researchers may have minimal say or responsibility about how their work is used. Geographers are increasingly valued for their ability to articulate complex human–environmental and scientific interactions to policy makers, even "inside the Beltway" (McMahon et al. 2005; Turner 2005; Roberson 2018). Scientists, stakeholders, decision-makers, and other key actors may find common ground for healthy debate on Pinelands land-change dynamics through geographic study of South Jersey's wetlands (Knight et al. 2008; Sanford et al. 2017).

Woodsman of the Pines

Early farmers had little interest in the Pinelands' impoverished gravelly and sandy soils. It was the timber stock, uniquely adapted to the region's inhospitable conditions (i.e., drought, frequent fires, nutrient-poor soils) that created economic stimulus to inhabit the Pines. Demand for lumber, naval stores, and charcoal was great during the colonial period (Sim & Weiss 1955; Weiss & Weiss 1965; 1968; Demitroff 2014), both for domestic use and for export to England. Forest resource extraction necessitated large teams of woodsman to fell, kiln, prepare, and haul the forest products to market. Mature wood was generally absent from the landscape from 1840–1940 (e.g., Cottrell 1937: 10; New Jersey Historical Records Survey Project 1940: 81)

To the woodsman, slight changes in the landscape were important to note. In a search for forest products, certain environments would prove better than others for exploitation. Vernacular, non-standardized names were coined for specific features (Lee 1896; Moonsammy et al. 1987: 13–41; Radis 1987), perhaps where water was plentiful, or cedar could be cut or game was plentiful. To the outsider, the distinctions between certain features would be unclear. Such is the case for localized hydrogeomorphic terms, such as *spungs*, *cripples*, *blue holes*, and *savannahs*. As noted by Chamberlain[2] (Bisbee 1971: 290), "People have grown old and died arguing as to the difference between a cripple and a spung." In a modern context, they are out of place. Their names are archaic, with their original roots long forgotten. The woodcutters, who originated and used these terms with a 'sense of place,' are now long gone.

Colonists quickly exploited Pine Barrens forest resources. The lumber industry began with the logging of Atlantic whitecedar (*Chamaecyparis thyoides* [L.]) in the seventeenth century to sheath homes in nearby

---

2. George Agnew Chamberlain (1879–1966) was an American diplomat and author who wrote short stories and novels, including some set in rural South Jersey (e.g., Chamberlain 1925). Amelia Earhart used to fly in and visit him on a regular basis (Hummel 2015).

Philadelphia (Thomas 1698: 27) and later New York City, although the virgin timber was for the most part felled by 1750 (Kalm 1770: 20). Not to be denied this valuable commodity, Pinelands woodsmen mined old fallen logs from the muddy, peaty swamp deposits (Weiss & Weiss 1965: 1–24). Other key forest products extracted from South Jersey's forests included charcoal (Simm & Weiss 1955; Gordon 2015) and naval stores such as pitch, tar, and turpentine (Weiss & Weiss 1965: 25–45; Wrench 2014). Forests were the coal mines and oil wells prior to the exploitation of fossil fuels. Carbon stored in wood fueled the iron furnaces (Pierce 1957; 1964; Johnson 2001), glass works (van Rensselaer & Boyer 1926; Pepper 1971), and paper mills (Fowler & Herbert 1976; Dellomo 1977) that once dominated the industrial landscape of the Pines. Cordwood, lumber, and hoop poles used in barrel manufacture were also important commodities (Wilson 1953: 357–368; Nickles c. 1958).

Edaphics: How the Sandy Soils have Influenced Living Things

Accounts of the first explorers to the New Jersey Pine Barrens describe a parched and inhospitable wasteland. During his travels of 1722, the English physician and botanist John Fothergill writes of his "journey through the deserts" of southern New Jersey, and of the "few dark people" who called this wilderness home (Fothergill 1754). Seven generations later, Platt (1889: 5) echoed Fothergill's sentiments when he noted how the Pine Belt's dry, sandy, sterile soils restricted its inhabitants to a most marginal existence. He proclaimed that the local person's "home is a hovel, and his food indigestible, and often insufficient and innutritious." In 1764, Wrangel (1969: 11) observed that the region's soils were primarily composed of drift sands[3]. Early maps (Fig. 1.2) denote this locale as a sandy barren desert (Evans 1749; Holland 1776; Adlum & Wallis 1791), an observation echoed in agricultural works. New Jersey Board of Agriculture (1893: 501, *also see* p. 62, "desert") reports, "I think this is the most God-forsaken country I ever saw; a crow would starve to death flying over it." Dillingham (1911: 106) states, "one gets the impression that the land is waste and desert."

---

3. 'Drift sands' is widely used in Europe to describe wind-moving sands for the European Sand Belt that blankets ice-marginal lowlands from Great Britain to Russia in drift-sand landscapes (Koster 2009).

Although precipitation is plentiful, averaging more than 1150 cm (45 inches) per year at Hammonton and Indian Mills (Ludlum 1983), the loose, sandy, and gravely soils allow rainwater to infiltrate quickly and recharge the shallow aquifer (Rhodehamel 1979b). Nutrients are leached away, leaving the soil impoverished and acidic. The soil, rather than the climate (i.e., the edaphics), makes the Pinelands drought prone. The flora and fauna are well adapted to the xeric conditions, and to the frequent forest fires that result (Fig. 1.3) (Little 1946; Boyd 1991: 15–17).

Shallow Groundwater

Groundwater in the Pine Barrens is never far below the surface. The Cohansey aquifer has excellent hydraulic characteristics and holds prodigious quantities of pristine water (Epstein 1985). Wherever the porous sands are eroded, wetlands and streams appear (Rhodehamel 1979b). Most stream flow is the result of groundwater seepage from the Cohansey aquifer. Surface runoff is negligible in the Pinelands (Vermeule 1894; Barksdale 1952; Rhodehamel 1979b). As with other streams and wetlands in the Pine Barrens, the numerous depressions and intermittent pools (i.e., pools that regularly become completely dry) derive their recharge by intersecting the water table (French & Demitroff 2001: 347), making them in some ways distinct from classical vernal pools. Pinelands ponds are almost entirely groundwater-replenished, whereas a typical vernal pool fills primarily from runoff, although often supplemented by groundwater (Kenney & Burne 2001; *also see* Parsekian & Slater 2011). The various water features are open systems, dependent on groundwater seeping through enclosing sandy and gravelly sediments with through flow from adjacent uplands, especially during the winter (Sun et al. 2006). They are usually not perched systems filled by the interception of runoff.

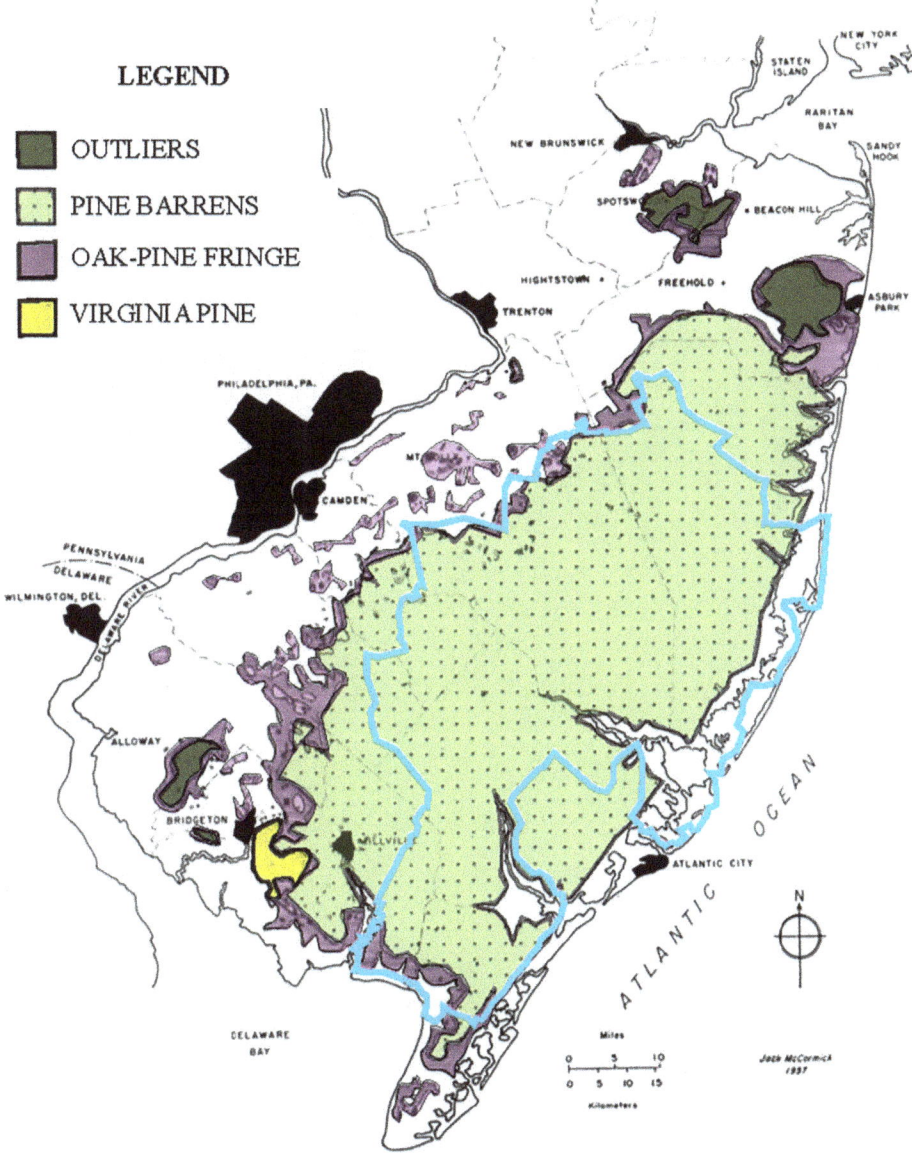

Fig. 1.1 Vegetational map outlining the boundaries of the Pine Barrens.

This map was adapted from McCormick and Andresen (1963: 23, Fig. 1). The Pinelands National Reserve is outlined in blue. Reproduced with permission from the New Jersey Audubon Society. The earliest iteration of this vegetational map is attributed to a map insert within Smith (1910), which shows Pine Barrens habitat once extended to Staten Island.

Fig. 1.2 Early map indicating desert-like Pine Barrens conditions.

The region was considered a "Sandy Barren Desert" (encircled) by early explorers, as depicted by Evans (1749).

Fig. 1.3 Representation of what early Pine Barrens forest looked like.

Circa 1900 photograph of a mature (old-growth?) stand of pitch pine (*Pinus rigida* [P. Mill.]) somewhere along or below South River, Weymouth Township, NJ (cf. Harper 1914: 363, park-like virgin forest of long-leaf pine). Little (1946: 3) imagined that frequent fires left little underbrush in the early Pine Barrens forest, and that travel by horseback or cart could be accomplished through tall stands of old bull pines without need of a trail. The forest canopy was imagined to be tight, which blocked light to understory vegetation. Frequent fires further discourage underbrush growth. This is thought to be a close representation of an original stand. Photo courtesy John Madara Collection.

02
Pine Barrens Geology and Geomorphology

Geologic Framework

The Cohansey and Bridgeton Formations underlie much of the Pine Barrens surface. Deposited during the middle to late Miocene (~15–5 ma), they consist of marine, marginal-marine, estuarine, deltaic, and fluvial deposits. Sands and gravels are common, but smaller deposits of silt, clay, cobbles, and boulders are also present (Newell et al. 2000). Since their deposition, the beds have undergone weathering, and exhibit the effects of prolonged leaching (Owens et al. 1983: 35). Salisbury (1898: 163–164), Tedrow (1986: 22), and French & Demitroff (2012) stated that these surfaces have been modified by slopewash and eolian activity. An overview of the region's Cenozoic history can be found in Stanford (2003) and Newell et al. (2000). The former paper deemphasizes the role of cold-climate processes in sculpting the landscape, attributing the current landscape to fluvial incision and erosion. The latter work suggests that the Coastal Plain is a badlands landscape resulting from periglacial mass wastage assigned to sustained cold periods during the Pleistocene.

Contested Pleistocene Inheritance

To appreciate the peculiar configuration of spungs, cripples, blue holes, and savannahs, background on the nature and recent modification of surficial deposits is necessary. Over most of the twentieth century, a spirited debate was waged about the recent geologic history of southern New Jersey (McCormick 1970: 83; French et al. 2003: 259–260, 2005: 2–3,

10–12; Stanford 2003: 44) involving such questions as: a) just how cold did this region get; b) were the frigid stages sustained or short lived; c) were cold periods wet and icy, or arid in nature; d) was the vegetation forest, tundra, taiga, or desert-like; and e) was permafrost ever present? With this facet of the region's geologic puzzle in place, a long overdue account of the Pinelands' distinctive wetlands can be advanced.

As to broader North American studies on Pleistocene conditions,[1] Smith (1949: 1497) stated, "As early as 1881 . . . Kerr[2] visualized the far reaching effects of climatic changes associated with glacial stages and postulated that the character and disposition of certain superficial deposits in the southern States were a product of frost action during the ice age." He added further that Chamberlain (1897: 826) remarked, "[it] must be remembered that as a result of the excessive superficial thawing and freezing incident to glacier-border conditions, the facilities for landslides, bodily creeps, and similar modes of movement reached an extraordinary degree of development." Smith (1949) also noted that neither of these two early observations prompted further investigation.

To describe the cold environment found just beyond the ice sheet, Walery von Lozinski (1909), a Polish geologist, first used the term "periglacial" as part of a study of rock-weathering conditions present during the Pleistocene in the Carpathian Mountains. His periglacial zone was adjacent to ice sheets and glaciers (French 2018: 5–8), with conditions that differed from current polar regions as a result of its mid latitude location. According to French (2000: 35), the term has since been broadened to include all cold, nonglacial environments "in which frost-related processes and/or permafrost are either dominant or characteristic." Periglacial conditions, as envisioned by Lozinski, were cold and arid, and with few modern analogs (French 2000; Ballantyne 2018: 7–9). It is well known that the Pine Barrens were 20–150 km (15–95 miles) south of the

---

1. A nineteenth century poet was the first to speculate that the Pine Barrens distinctive landscape was a reflection of Ice Age forces (Hopping 1886). In an obscure monograph called *Reminiscences of Atlantic County: Mays Landing*, there is mention of a time when South Jersey consisted of treeless plains and broad shallow valleys. To the poet, the local relief was recently "denuded" by the "long winter," caused by the shifting of the Earth's balance or "divergences of polarity in its long journey through space" (Hopping 1886: 10). He followed this "long winter" with the "torrid age of the Dinosauria." Where Hopping received his illumination for such a modern insight remains unknown, but it suggests that informal discussion over South Jersey's unique landscape was taking place well before explanations appeared in scientific literature.

2. (Kerr 1881).

Late Pleistocene ice sheet margins (Fig. 2.1), and so experienced cold, nonglacial conditions (Péwé 1983; Newell et al. 2000; French & Demitroff 2001; Merritts & Rhanis 2022).

Salisbury (1893) identified ventifacts (wind-faceted stones) in South Jersey. The presence of ventifacts is indicative of hot or cold desert-like conditions (Laity 1994: 507). Later, Salisbury and Knapp (1917: Figs. 7, 21, 41) recorded unusual features in the Quaternary sediments of southern New Jersey but did not ascribe a geomorphic explanation to them. These features now could be interpreted as periglacial in nature (Salisbury & Knapp 1917: Figs. 21, 41), including relict thermal contraction wedges (Fig. 2.2), modified frost cracks (Fig. 2.3), and an apparent shear fault that could have only occurred in frozen sediments. The chaotic gravels noted in delightful fashion by Salisbury and Knapp (1917: Fig. 31) as "promiscuous" in their distribution, are likely frost-disturbed in nature. Various structural features were captured on photographs by Salisbury and Knapp that are remarkably similar to those attributed to intense frost action in the Pine Barrens (French & Demitroff 2001, 2003; French et al. 2003, 2005, 2007, Demitroff 2016).

Wolfe (1941) was the first to conjecture that Late Pleistocene periglacial processes acted upon the surficial deposits beyond New Jersey's glacial moraines. He described how the valley types, *dales* and *vales*, were probably formed under Pleistocene conditions, when vegetation was sparse and wind and fluvial erosion were severe (Wolfe 1941: 55–56, 1943: 209). Wolfe speculated that dune fields might once have covered the Frenchtown area, about 70 km (44 miles) northeast of the Pine Barrens (Wolfe 1941: 25– 26).

Wolfe (1952; 1953; 1956) later proposed that permafrost had existed in the Pine Barrens during the Late Pleistocene. For glacial periods, he envisioned a forested region, with Spruce (*Picea*), Fir (*Abies*), and Hemlock (*Tsuga*). Low-lying areas were to be ponded, with extensive peat deposits. In places, the organic mats would have attained a thickness of three meters. In Wolfe's "frost-thaw" model, much of southern New Jersey was underlain by permafrost containing ice wedges and thick lenses of segregated ground ice. Wolfe also noted signs of strong wind action. He observed dunes in sandy areas throughout the Coastal Plain, widespread wind-blown loess[3] deposits, and numerous ventifacts (Nichols 1953).

---

3. Loess in simple terms is windblown silt, which was deposited on New Jersey's Inner Coastal Plain along the downwind edges of the Delaware River—but attenuating in sedimentation before reaching the Pine Barrens (Newell et al. 2000; Fig. 3; Wah et al. 2018:

The wide extant of eolian material did not support Wolfe's concept of a cold, humid, peat-filled, and forested Pleistocene landscape, yet he made no attempt to explain away the implication of this extensive wind-action. Doubt was also cast upon his supposed periglacial features: the "frost-wedge fillings" on Plates 2B and 2C (Wolfe 1953, 137) were not convincing ice-wedge casts (Brown and Péwé 1973: 91; Péwé 1973: 22; 1983: 166). To Péwé (1983: 179), the complex deformation structures (Wolfe 1953: Plates 2A & 3A) described were possibly suggestive of the prior existence of perennially frozen ground but did not prove it. Rasmussen (1953: 473–474) accepted Wolfe's premise that New Jersey experienced a periglacial climate during the past but felt that insufficient evidence had been provided to link periglacial processes with the origin of his enclosed basins. Pierzchalko (1954: 99) was troubled that Wolfe's wedges were not found in direct association with basin sections and questioned their freeze-thaw genesis. He credits Wolfe's paper as the first through fieldwork to invoke Pleistocene ice thaw (thermokarst) for basin genesis.

Wolfe (1977: 163–172, 290–293) revisited the problem of permafrost in the Pine Barrens in his textbook. Although reiterating earlier work, he added some macrofaunal fossil evidence for periglacial conditions, and patterned ground phenomena as reported by Walters (1975). Notably, Wolfe stated with some certitude that there had been Pleistocene permafrost. He concluded that the "Coastal Plain of New Jersey was a treeless windswept tundra plain during the Pleistocene Epoch—much like that of Siberia today" (Wolfe 1977: 163).

Another early reflection on periglacial conditions was given by Hack (1953: 186) who asserted, "A zone of intense frost action borders the Wisconsin ice sheet in . . . New Jersey, analogous to Büdel's frost debris tundra." To Hack's vision of periglacial New Jersey, Late Pleistocene conditions began with a cold moist climate, followed by a cold dry climate. As evidence, Hack recounted a conference hosted by the "Friends of Pleistocene Geology" in 1949 where P. MacClintock showed attendees "involutions, ventifacts, and other evidence of cold climate" (Hack 1953: 186). Periglacial features were presented on this field trip at several locations (Lantis 1949: 10) including Fairless Hills, Pennsylvania,

---

788, Fig. 1). Some silt was noted in association with Pinelands paleodunes (Demitroff et al. 2019). While there is uncertainty as to loess' origin and paleoenvironmental significance, Pleistocene loess is associated with cold, windy, and arid conditions (Ballantyne 2018: 291–294). Konishchev and Rogov (2017) suggest that cryogenic weathering under very harsh climate conditions could be a prime source of loess production.

and Dayton, New Jersey (personal communication, John Tedrow, July 2005).

More evidence for a periglacial New Jersey was published during this period. Tedrow and MacClintock (1953) reported a band of Pleistocene loess perched atop shale deposits in the vicinity of Trenton. Some of the loess was supposedly buried by cryoturbation from alternate freezing and thawing. Nearby, on the sandy and gravely coastal plain formation, the loess had degraded beyond recognition because of frost heaving and "turbation." Tedrow and MacClintock cited Harper (1950) for documentation. The loess on the Coastal Plain was mixed by "frost turbulence" during the "tundra episode of the Pleistocene." Tedrow (1963: 7) later extended the loessal zone, expanding the influence of eolian activity throughout the New Jersey Coastal Plain. This wind action was inferred from laboratory and field evidence.

As State Geologist, Widmer (1964) authored a review of the Garden State's geology in celebration of the statehood's Tercentenary. A popular work, not intended to be a defendable scientific endeavor, the study was designed to "permit the interested amateur to understand the complexities of the geology and geography of New Jersey, without getting lost in the technical details" (Widmer 1964: ix). To Widmer (1964: 136–138), the New Jersey Coastal Plain was a cold, arid, and windy desert during periods of the Pleistocene. He alluded to new evidence of relict material in New Jersey, such as ice-wedge polygons, ice cracks, widespread loess deposits, desert pavements, sand dunes, and frost weathering. According to Widmer, these were evidence that a very cold climate developed in the region beyond the ice sheets.

McCormick authored "A Study of Significance of the Pine Barrens of New Jersey" (1967, reissued as McCormick 1970). His investigation was directed by the Academy of Natural Sciences in Philadelphia and drafted for the National Park Service. Its purpose was to advocate the numerous ecological research opportunities present in the Pine Barrens. McCormick's report laid the groundwork for the Pinelands National Reserve and became a rallying point for many ecological, botanical, and zoological investigations. His frustration over the varying interpretations of the region's Pleistocene phenomena was apparent. McCormick (1970: 83) acknowledged, "Some geologists claim to have found evidence of severe tundra-like conditions, but others claim the region was subjected to conditions only slightly cooler and wetter than at present." Recognizing outstanding potential for geological study, he suggested that the Pine Barrens was an advantageous place for researchers because of its proximity to academic and governmental institutions, its pristine state, and its simple terrain.

The Frozen Landscape of Newell and Others

Expounding upon the "harsh periglacial"[4] conceptions of Wolfe (1953, 1956) and Widmer (1964)—also Newell and Wyckoff (1992)—surmised that the paleodrainage patterns of four Coastal Plain watersheds had all the characteristics of stream channels in areas where permafrost[5] is present today. They argued that thawing and dewatering of the top meter of sediment resulted in widespread debris flows. The two geologists gave an account of cold-climate features they observed, such as braided streams, fossil frost wedges, cryoturbation, ventifact pavements, and windblown sands—all of which were "signatures of [perennially] frozen ground" (Newell & Wyckoff 1992: 27).

Newell et al. (1989) touted the "well developed periglacial features" along the route of a field conference in November 1988. Mention of past permafrost is noticeably absent from the field guide, despite numerous references to frost wedges and cryoturbation. These authors cautiously attributed the cold-climate phenomena to "frost action" and "freeze and thaw." They were careful not to postulate the past occurrence of perennially frozen ground, as it was the proverbial "third rail" to the East Coast geomorphological community (personal communication, Wayne Newell, August 13, 2018).

Nearly a decade passed before Newell again wrote about southern New Jersey's geology. Newell et al. (2000: frontispiece, 5) concluded that permafrost had been present in the Pine Barrens during the Late Pleistocene. As in an earlier publication (Newell & Wyckoff 1992), it was the nature of the paleohydrology that provided the best evidence of past perennially frozen ground. Newell et al. (2000) credited cold-climate processes as potent geomorphological agents in the landscape evolution of the New Jersey Pine Barrens. They stated that there is "evidence of cold, continental climates and periglacial phenomena . . . abundantly preserved on the New Jersey Coastal Plain," and that this cold-climate

---

4. Originally periglacial meant "bordering glaciers" but is now extended to all cold, nonglacial environments (Ballantyne 2018: 1).

5. Permafrost refers to ground in which the temperature remains below 0°C (32°F) for two years or more (Ballantyne 2018: 1).

period was sustained (Newell et al. 2000: 9). Periglacial conditions were inferred from the widespread presence of relict frost wedges, cryoturbation, periglacial mass wasting, and other slope processes, eolian sands and loess, braided streams, and "thermokarst" ponds.[6] In recent reports, periglacial slope processes were explored inside (French et al. 2007) and just outside the Pine Barrens (Newell 2005; Newell & DeJong 2011).

Palynological Evidence

Pollen data were often at odds with the geological evidence (Demitroff 2016: 123–124). Palynological studies seemed to undermine the theory of a protracted periglacial period occurring during the Wisconsinan (e.g., Custer 1986; Stanford 2003). Pollen analysis from southeastern Pennsylvania and northern Delaware (e.g., Watts 1979; 1983; Russell & Stanford 2000) suggested that open spruce/fir woodland predominated in southern New Jersey during the Late Wisconsinan, with a narrow band of tundra proximal to the ice sheet. Custer (1986: 22) indicated, "The available pollen data from the Late Glacial and early Holocene timeframe in the mid-Atlantic region do not match any of those associated with periglacial bogs." Groot et al. (1995) suggested a temperature shift of about 10°C occurred during Late Pleistocene cold periods.

Caution should be exercised when looking at the palynological data. Pollen dating to the glacial maximum period between ~27 and ~14 ka,

---

6. Others commented briefly on probable deep seasonal frost features on the New Jersey Coastal Plain. Those who remarked on such matters included: 1) Nichols (1953)—ventifacts, involutions; 2) Martens (1956: 75)—cryoturbation; 3) Richards and Rhodehamel (1965: 13)—cryoturbation and frost heaving at Cape May; 4) Richards and Judson (1965)—freeze-thaw basins, frost-affected soils, frost wedge, ventifacts, loess and wind-blown sands; 5) Bowman (1966: 22, 23)—periglacial deformations to a twelve foot depth; 6) Bowman and Lodding (1969: 5)—cryoturbation; 7) Minard and Rhodehamel (1969: 286, 300)—cryoturbation, ventifacts, modified frost wedges; 8) Tedrow (1973: 292, Plate 2)—a wedge-like feature; 9) Walters (1975)—ice-wedge casts, patterned ground, cryoturbation, ventifacts, asymmetrical valleys; 10) Walters (1978)—ice-wedge casts, patterned ground, ventifacts; 11) Tedrow (1979, Fig. 6, right side), wedge-like structure; 12) Rhodehamel (1979a)—ventifact cobbles and boulders; 13)—Bonfiglio and Cresson (1982)—pingo and palsa scars; 14) Trela (1984)—frost action on soils, cryogenic hardpan; 15) Tedrow (1986)—wedge structures, sand dunes, loess, ventifacts, periglacial basins, wavy ironstone, sag structures, vesicular and fragic soil micromorphology; 16) Stanford (1997: 6–17)—open-system pingos, thermokarst basins; and 17) Stanford (2003: 36, 38)—thermokarst basins, cryoturbation, scattered windblown sands.

28   Soggy Ground

has not yet been recovered from southern New Jersey (Stanford 2003: 31–32, Table 1), although a new study by Arsenault (2022) near Clayton, New Jersey—just outside the Pine Barrens—holds promise to add new insight to this problem. A series of continuous spung sediment cores were radiocarbon dated and preliminarily appear to provide a depositional record that spans from the Late Glacial Maximum to the present day. The conventional wisdom is that pollen had been deposited but had degraded due to environmental conditions (personal communication, Scott Stanford, February 2004). Alternatively, the absence of datable material could be explained by Late Wisconsinan conditions that were too extreme for accumulation of pollen-preserving sediments or peat deposits (Hartzog 1982: 7).

Renewed Interest in the Periglacial Environment

The combined work of French et al. (2003, 2005, 2007) and Demitroff (2016) described and assigned dates to relict sand wedges (Fig. 2.2) from the Pine Barrens, validating the prior existence of the sustained periglacial environment proposed by Wolfe (1953, 1956, 1977); and Newell and Wyckoff (1992); Newell et al. (2000). Frost wedges described from Delaware (Lemcke & Nelson 2003; 2004; French et al. 2009a; Losco et al. 2010), Maryland (French et al. 2009a); and Pennsylvania (Gross et al. 2017; Merritts et al. in review) indicate that periglacial conditions were widespread, as does thermal-contraction polygon traces on Google Earth imagery of New Jersey, Delaware (Gao 2014), and Pennsylvania (Merritts et al. 2015; Merritts & Rhanis 2022). These studies related attributes of the region's many deformation structures (Figs. 2.3–2.7) to degradation of permafrost.[7] The deformation structures were likely modified by density readjustment in water-saturated sediments, and by fluvio-thermal erosion. Another effect of the melting of icy beds is seen in broken and disrupted ironstone layers (French et al. 2003), believed to have 'foundered' in sediments that lost cohesion through fluidization or liquefaction of sediments when permafrost thawed (Fig. 2.8).

---

7. A common misconception is that *permafrost* and *ground ice* are synonymous (French 2002). Permafrost is a condition based on time and temperature. Succinctly stated by van Everdingen (1998: 54), "It [permafrost] is not necessarily frozen, because the freezing point of the included water may be depressed several degrees below 0°C; moisture in the form of water or ice may or may not be present." In other words, whereas all perennially *frozen ground* is permafrost, not all permafrost is perennially frozen.

Much of the regolith in South Jersey was likely to have been ice-poor, judging by the nature of regional periglacial macrostructures. Small sand-filled fissures found within an argillic truncated B horizon (periglacial fragipan) are interpreted as desiccation cracks (French & Demitroff 2003; et al. 2005; Demitroff 2016; cf. Washburn 1979). The type of ground-wedge casts described by French et al. (2003, 2005) and Demitroff (2016) is often attributed to cryodesiccation (Paepe & Pissart 1969; Pissart 1970; van Vliet-Lanoë 1983) or convection (Gubin & Lupachev 2008) or hyperarid drying and wetting (Li et al. 2014) under deep seasonal frost or in the active layer above permafrost or of multiple derivations (Wolfe et al. 2018). Widespread sand wedges (French & Demitroff 2001; et al. 2003; 2005; Lemcke & Nelson 2003; 2004; Newell 2005; & DeJong 2011; Demitroff 2016) result from thermal contraction in cold sandy, often desert-like environments (Murton, et al. 2000; Worsley 2016: 16; Ballantyne 2018: 96; French 2018: 153) with little or no moisture available to enter the crack. Soils adjacent to South Jersey's ground, sand, and deformed wedges, as well as soils enclosing sediment-filled pots, are not deeply deformed, indicating that permafrost was ice-poor (Demitroff 2016). Van Vliet-Lanoë (1996: 462) reached similar conclusions about ancient wedges in France, Belgium, the Netherlands, Germany, and Poland. In New Jersey, as in northwestern Europe, wedge modification by thermokarst processes was slow and simple (Fig. 2.6), with deformation occurring mostly by running water in summer creating sand-filled gullies (French et al. 2005; Demitroff 2016). The widespread distribution of ventifacts (Figs. 2.9–2.12) and dunes (covered in later chapters) also indicates that dry conditions were prevalent (Demitroff 2007—dunes; 2010—wind streak; 2016—dunes, ventifacts; Demitroff et al. 2012—dunes; 2019—dunes; French and Demitroff 2012—dunes, ventifacts; Markowich et al. 2015—dunes; Swezey 2020—dunes; Wolfe et al. 2023—dunes, ventifacts).

It was concluded that perennially frozen ground existed in the Pine Barrens during at least two periods (French et al. 2003, 2005), and more likely three or more (Demitroff 2016).[8] These periods correspond to

---

8. On July 12, 2006, an informal United States Permafrost Association field trip, led by Demitroff, Nelson, and Newell, was held in association with the World Congress of Soil Science, and visited several of the locations investigated by French et al. Arctic and Antarctic cryosol authorities on this trip interpreted macro- and microstructures shown as periglacial in origin. Although the exact nature of permafrost remains ambiguous, the observed lenticular platy structures and associated sand wedges (> 2 m in depth), along with other attendant microfabrics and macrostructures (Huijzer 1993), were convincing evidence to the attendees that perennially frozen ground was once present in southern

Marine Isotope Stage (MIS) 6 (~200–140 ka), MIS 4 (~70–50 ka), and MIS 2 (~17–13 ka) determined by optically stimulated luminescence (OSL) dating methods (Figs. 2.13–2.15) (French et al. 2007; Demitroff 2016). Perennially frozen ground may have extended as far south of the Pine Barrens as North Carolina (French & Millar 2014: 3, Fig. 1; Lindgren et al. 2016: 9, Fig. 2; Marshall et al. 2021: 48, Fig. 2) (Fig. 2.16). At times the permafrost may have been patchy during the Late Wisconsinan with denuded uplands subject to strong wind action, and lowlands occupied by shrub–forest and tundra with some snow cover. It appears that earlier very cold periods were semi- to desert-like, with a mean annual air temperature (MAAT) -8 to -6°C (Early Wisconsinan and/or Illinoisan; cf. Groot et al. 1995, MAAT 2 to 3°C), or provisionally even colder (Demitroff et al. 2007: 141, -12 to -10°C), was more sustained, and permafrost may have underlain the entire landscape (French et al. 2005: 173, -4 to -3°C; 2009a: 286 <-6C; 2009b: 27, -8 to -6°C; Marshall et al. 2021, -12°C at the ice front) (Fig. 2.13).

Epilog to the Spirited Debate

In conclusion, there is strong evidence that the Pine Barrens experienced periods of periglacial conditions. Ancient sand wedges, thermokarst structures, windblown sediments, and other phenomena are well preserved in southern New Jersey. The region was affected by strong katabatic winds from the continental ice sheet to the north. Perennially frozen ground, accompanied by sparse vegetation, episodically characterized the environment. This landscape may be analogous to parts of Europe during the Pleistocene (Vandenberghe et al. 2014), where geomorphologists have scrutinized evidence of periglacial environments for nearly a century (Lozinski 1909; Cailleux 1936; Büdel 1944; Dylik 1952; Boardman 1987). It is likely that Late Pleistocene cold epochs left their mark on the spungs, cripples, blue holes, and savannahs, and that cryogenic processes were active agents in landform genesis. This work attempts to locally uncover geographic elements of this periglacial legacy.

New Jersey. Field trip attendees included G. Broll, C.-L. Ping, C. Tarnocai, F. Ugolini, E. Ciolkosz, H-P Blüme, S. Buchanan, A. Benthem, D. Konyushkov, and A. Makeev.

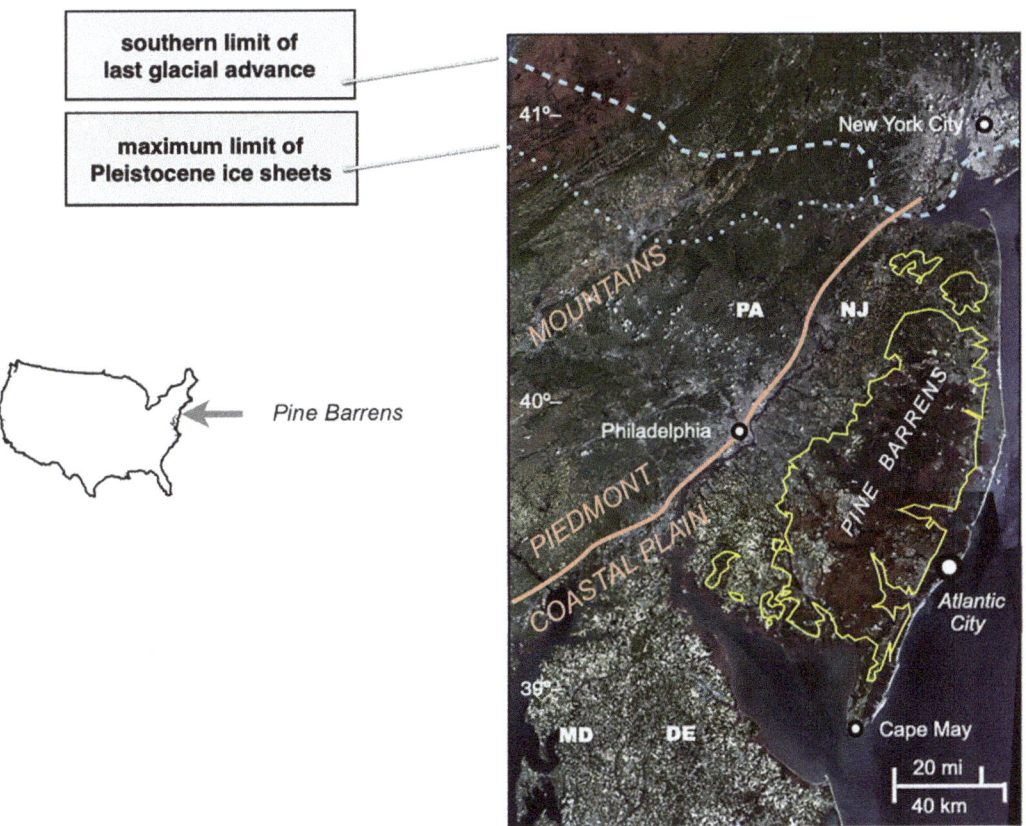

Fig. 2.1 Satellite map of Pine Barrens ice marginal position.

The New Jersey Pine Barrens is a distinctive place, even from space. Note the spectral color reflectance that can be seen in satellite imagery. Yellow lines used to bound the Pines in this diagram are adapted from McCormick and Andresen (1963: 23, Fig. 1) (Fig. 1.1). NASA LANDSAT, late March 'True Color' 3, 2, 1 band combination photographs L3-L71014032_03220020510_B30 and L3-L71014033_03320020424.

Fig. 2.2 Periglacial wedge-related structures identified in the region.

(A) Photo of an older sand-wedge cast, optically stimulated luminescence (OSL) dated at greater than 65 ka (French et al. 2003, 2005). Older wedges in the Pine Barrens are often filled with laminae (sand-vein layers) of fine, highly wind-abraded sand grains that became moderately densified with time. The frost-crack's depth (2.7 m) suggests that widespread permafrost was present during its genesis (Murton et al. 2000; Christiansen et al. 2016; Worsley 2016; Merritts et al. 2022: 549, Table 2), thus bringing the realm of continuous perennially frozen ground to the latitude of Baltimore, MD, and Washington, DC.

(B) A thin primary sand-vein offshoot (Murton et al. 2000: 900, wedge apophyses; Merritts & Rhanis 2022: 564, Fig. 8) is indicated by the trowel. Dorchester pit, near Port Elizabeth, NJ. Photomicrographs by SEM Laboratory, Department of Earth Sciences, Carleton University.

Several types of frost wedges are referred to in this work. All are inoperative under current warm interglacial conditions we now experience. That makes the fissures ancient, relict, or paleo- features. Herein, the convention is to use "cast" for all frost-wedge features like ice-wedge, sand-wedge, ground-wedge, or composite-wedge casts. Cast has several definitions, one paleontological and another sedimentological. However, casting in the purest sense means the simple action of throwing (flinging, hurling, pitching, tossing) something (OED June 2022).

- **Sand-wedge cast** — wind-lifted infill is thrown or casted into a crack
- **Ice-wedge cast** — slumping host material is cast into the void left by melting ice.
- **Ground-wedge cast** — wind-lifted and slump-produced infill is cast into a void.
- **Composite-wedge cast** — wind-lifted and slump-produced infill is cast into a void.

A simple Google search (July 06, 2022) returned these results: "relict ice wedges" — 1820; "ancient ice wedges" — 2,240; "ice-wedge pseudomorphs" — 5,330; "ice-wedge casts" — 15,900; "sand wedge pseudomorphs" — 5; "ancient sand wedges" — 494; "sand-wedge casts" 2,420; and "relict sand wedges" — 3,530.

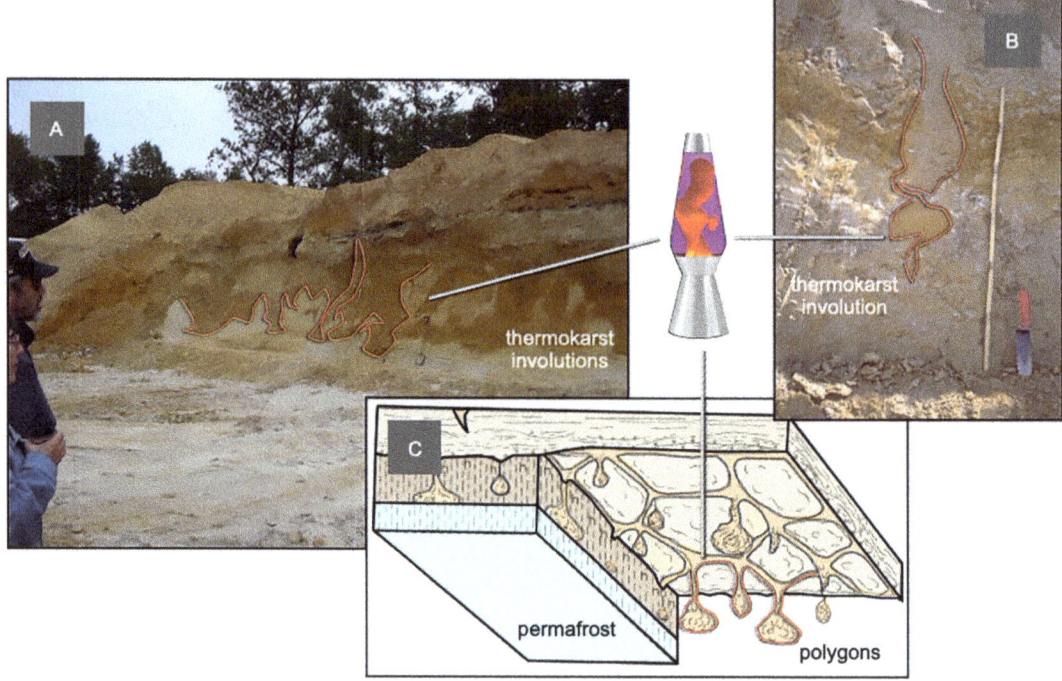

Fig. 2.3 Periglacial thaw-related structures identified in the region.

(A) These complex deformations are interpreted as thermokarst involutions (French et al. 2003, 2005). In the Pine Barrens, thermokarst involutions are widespread features where ground amorphously deformed with permafrost thaw ice melt. A lava-lamp-like density readjustment occurred through mass displacement as the thaw zone became increasingly saturated with perching of meltwater above a cryogenic aquiclude or aquitard (Anketell et al. 1970; Murton & French 1993), Unexpected pit, Newtonville, NJ.
(B) Permafrost-thaw deformation at a thermal-contraction crack intersection in fine sands, silts, and clays. Horsey pit, near Laurel, DE.
(C) Vandenberghe & van den Broek (1982: 306, Fig. 8) noted similar wedge-intersection involutions in clay near Meerle, Belgium. Even today, ancient frost cracks translocate water in a manner to agricultural tile drains (Woronko et al. 2022b, sand-wedge-affiliated geochemical corridors). Reproduced with permission from ©John Wiley.

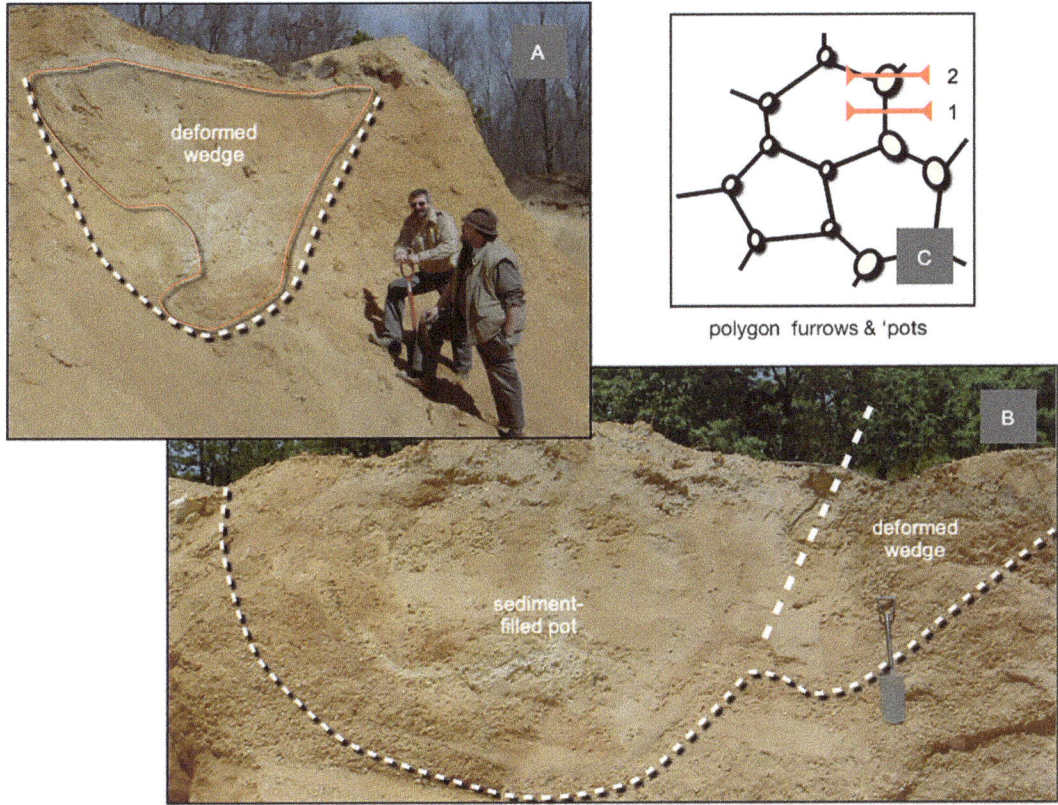

Fig. 2.4 Thaw-related sediment-filled pot and deformed wedges.

Unexpected pit, Newtonville, NJ.

(A) A deformed sand wedge, outlined in red, was reworked through thaw erosion, gullying, and slumping. Such a large trough would have acted as a line of preferential thaw in perennially frozen ground (French et al. 2005).
(B) This modified wedge was situated adjacent to the juncture of several frost cracks, which all combined to form a sediment-filled pot. White, powdery ash-like material from within the wedge (Demitroff et al. 2007).
(C) Past presence of frozen ground is often accompanied by polygonal patterns, generating 8–40 m polygons made by wedge furrows (see Gao 2014; Merritts & Rahnis 2022; et al. in review). Line (1) marks the approximate position of A's deformed wedge and Line (2) marks the approximate position of B's sediment-filled pot in relation to other furrows and pots.

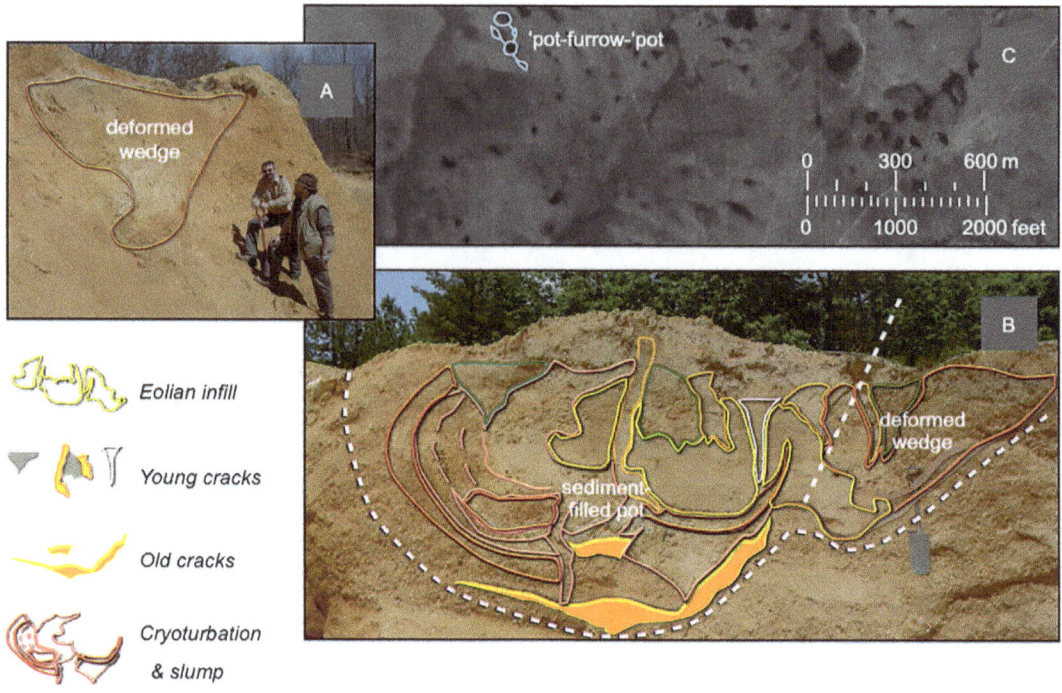

Fig. 2.5 Deformed wedges, sediment-filled pots, and 'pot–furrow–'pot topography.

(A) Deformed wedges accommodate complex infill that apparently relate to several cycles of permafrost aggradation and degradation. Their concave forms often reside in densified host sands, gravels, and silts or fragipan, which aided in their ancient form fossilization during permafrost thaw (thermokarst). Unexpected pit, Newtonville, NJ.
(B) Sediment-filled pots form as thermokarst bulbs at deformed wedge intersections. They too accommodate complex polycyclic infill. Unexpected pit, Newtonville, NJ.
(C) Buried wedge and sediment-filled pot remains were excavated by wind erosion and slopewash, providing the template for the distinctive 'pot–furrow–'pot topography associated with small spungs in the Pine Barrens (Parsons Ponds, Canute Neck, Vineland, NJ) and beyond (PA—Poster et al. 2013: 103, Fig. 2; MA—Calhoun & deMaynadier 2008, Plate 1). Aerial imagery (NJOIT-OGIS c. 1931, accessed through 1930's Aerial Photography of New Jersey, Boyd Ostroff).

Fig. 2.6 Deformed wedges and gullies.

(A) Modified or deformed sand or ice wedges—and their intermixed 'composite wedge' forms—might become wide linear gullies during climate-warming thermokarst. In this case, an older (≥ MIS 4, Early Wisconsinan— ~71–57 ka*) relict sand wedge (rat-tail left and below trowel) formed along a linear track beneath the modified wedge carapace formed of densified sands, silts, and gravels (outlined in red) which ran parallel with the underlying relict sand wedge. With MIS 3 (Mid Wisconsinan, ~57–27 ka) climate amelioration the sand wedge became increasingly wet with permafrost thaw ice melt and a return of greater atmospheric precipitation. With thaw, the previously frozen sediments (esp. sand) could now be picked up by the wind from the surrounding sparsely vegetated area and infilled the gully (OSL dated at red star to 34,800 years ago). With MIS 2 (Late Wisconsinan— ~27–11.7 ka) cooling permafrost returned, initiating a sand wedge generated (OSL dated at yellow star to 16,800 years ago) in the now-frozen MIS 3 sandsheet sand. Unexpected pit, Newtonville, NJ.
(B) Nichols (1969: 73, Figure 34 shown) reports a sand-infilled modified wedge frost-furrow structure on vegetation-free alluvial fans in Dallas Bugt, northwestern

Greenland, which may be a modern analog for structure A. Reproduced with permissions from Museum Tusculanum Press and the Department of Earth and Climate Sciences at Tufts University.

*A good argument is made by Otvos (2015) that the Early Wisconsinan should be extended back to ~115 ka, which this author fully concurs with (also see Merritts & Rahnis (2022: 549–550).

Fig. 2.7 Paleoperiglacial structures can be difficult to see in exposures.

It is easy to overlook ice age features without tedious and meticulous preparation beforehand. Unexpected pit, Newtonville, NJ.

(A) This wall shows thermokarst involutions on display during a field excursion for the Twentieth Annual Meeting of the Geological Association of New Jersey, October 10–11, 2003, hosted by Richard Stockton College of NJ (French & Demitroff 2003: 138–140; also see French et al. 2005: 177, Fig. 3). It took French and Demitroff over a week to properly prepare the site.
(B) The author revisited the same pit face in anticipation of a World Congress of Soil Science field excursion in 2006. The exposure looked uninteresting before re-excavation.
(C) Periglacial features revealed themselves only after careful pit-wall preparation.

Fig. 2.8 Ironstone boulders founder during thermokarst.

(A) In the NJ Pine Barrens, ironstone blocks within 10–15 meters of the surface can be found in a tilted position. In French et al. (2003) such boulders are interpreted as having "foundered" when permafrost thawed. Boulder tilting during thaw would be an artifact of density readjustment in super-saturated unconsolidated 'soup' conditions. Such tilting would be limited to environments as might be found above a frozen aquiclude or aquitard in a river channel. Moisture sufficient for this readjustment to take place might be found in association with a riverbed talik, but not so in uplands under local past permafrost conditions as we understand them. Tilted blocks can also result from river icings. Alekseyev (2015: 183) states, "In spring, the heaving mound [of aufeis] was destroyed, and the rock slabs were literally broken out of the rock mass and set at an angle of about 40 degrees against their former position." Southern Industrial Sands, Folsom, NJ.
(B) Ironstone is usually formed in wet areas as a planar bed at a groundwater table catalyzed by bacteria (Means et al. 1981), especially savannah wetlands. Ore Pond, near Chatsworth, NJ Photo taken by Jeff Larson.
(C) Here at Buck Pit ironstone was uncovered at great depth (+20 m) with its planar bedding preserved. It was unaffected by thermokarst tilting. This horizontal slab was buried under many meters of sediment, too deep to be affected by thermokarst processes. Pine Hill, Perrineville, NJ.

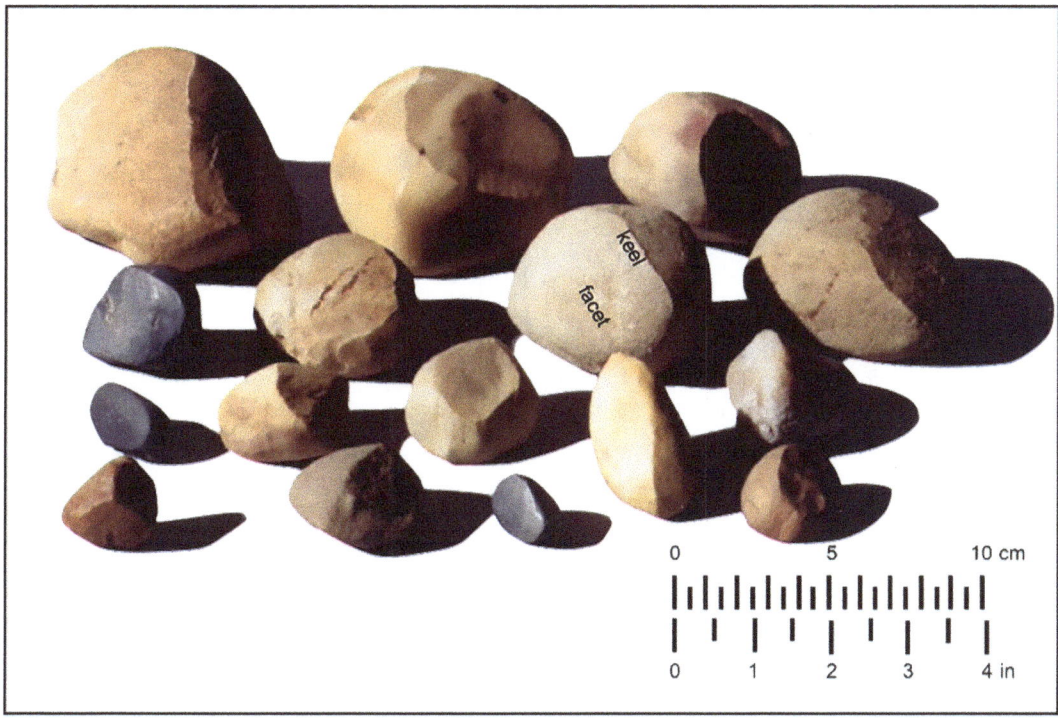

Fig. 2.9 Gravel-sized ventifacts.

Wind-abraded gravels are common in the Pine Barrens (Nichols 1953; French and Demitroff 2001, 2003; Demitroff 2016). Blowing sand is the primary abradant. Many southern New Jersey ventifacts are composed of quartz and quartzite. Collected in Cape May, Atlantic, and Cumberland counties, NJ. Adapted from photo taken by Mike Hogan.

Fig. 2.10 Cobble-sized ventifacts.

Wind–faceting and polishing of scattered surface-derived cobbles also occurs in the New Jersey Pine Barrens, Atlantic and Gloucester counties. While it is easy to invoke blowing sand as the abradant responsible for facet production, sand-grain's role in the formation of flutes and grooves is not understood (Demitroff 2016). Adapted from photos taken by Mike Hogan.

(A) Examples from Camden and Atlantic counties: (1) an orthoquartzite cobble with a single facet on the upwind side called an einkante; (2) a Brazil-nut shaped sandstone cobble with a single facet—an einkante—on the upwind side and with grooves carved into its downwind side; (3) an ironstone block coated with silica glaze with embedded quartzite pebbles worn flush along with the planar faceting of the block's upper surface; and (4) a sandstone cobble coated with silica glaze with a single facet—an einkante—on the upwind side and flutes carved into its downwind side.
(B) A close-up of ironstone block (3) showing wind-etching to embedded pebbles that are wind-cut flush to block surface. Specifically, a "kante" (from German) is the keel at facet's edge; "ein" is one keel, "dri" is three, etc.

Fig. 2.11 Boulder-sized ventifacts.

Wind–scalloping and polishing of ironstone boulders on Pine Hill, Perrineville, NJ, about 25 km south of the Laurentide Ice Sheet's maximum advance.

(A) Grooves, some helical, are deeply etched into the face of this boulder.

(B) Some boulders are dimpled with numerous weathering pits called opferkessel, indicated that have had a long exposure to the air.

(C) Some opferkessel have been wind-modified into wind-oriented coal-bucket shapes. Intense wind-etching is further evinced by the presence of flutes, grooves, and the terrestrially rare scallop form of ventifact (Minard 1966; Demitroff 2016).

Fig. 2.12 A wind–scalloped, grooved, and fluted sandstone boulder.

Although rare, large non-native boulders are on occasion identified in the Pines, here Vineland, NJ. Their exact origin remains a mystery. Perhaps the boulders are dropstones freed from floating river ice. Alternatively, the rivers that brought them here may have been energetic enough to carry such boulders in their currents.

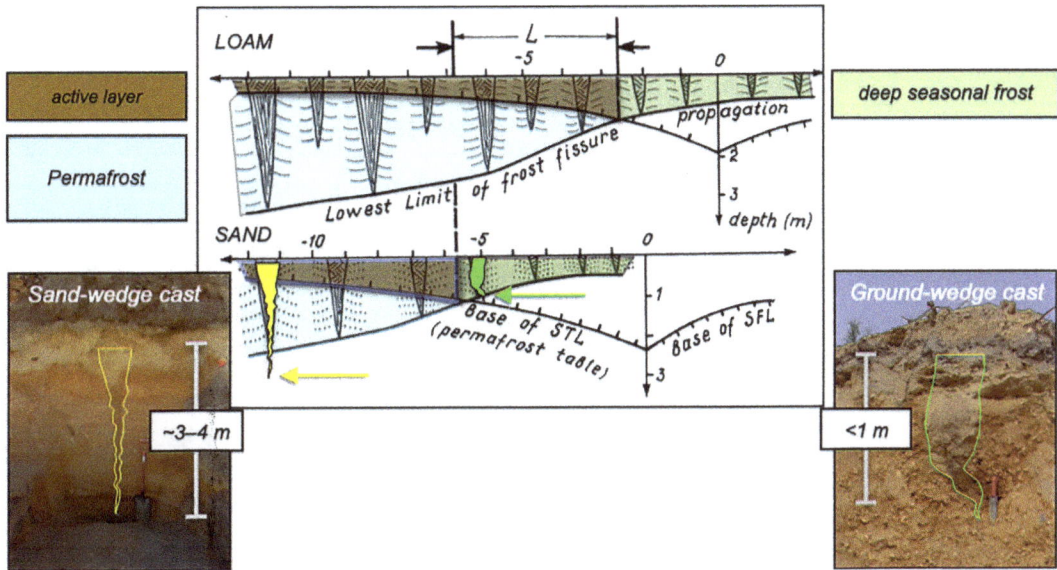

Fig. 2.13 Wedge-structure depth and hosting substrate indicate paleotemperatures.

Romanovskii's (1985: 159, Figure 9-3) model of the relationship between Mean Annual Air Temperature (MAAT), the cracking substrate nature, and frost-wedge formation. How applicable it is to midlatitude past permafrost is open to debate. Suffice it to say that it is easier to form wedges in silts than in sands due to the thermal dynamic qualities of each deposit. There are several types of frost wedges. For simplicity, we will concern ourselves with thermal contraction wedges—those developed in permafrost—and ground wedges—those formed in deep seasonal frost in the absence of permafrost.

On the bottom left, outlined in yellow, is a sand-wedge cast in Pine Barrens Bridgeton Formation that in Siberia would requires a MAAT of -12° C to form, although French et al. (2009b) publish a MAAT -8 to -6° C during the coldest Pine Barrens episodes. On the bottom right, outlined in green, the ground-wedge cast in Pine Barrens Bridgeton Formation that requires a MAAT of -5 to -4° C to initiate such cracking. Reproduced with permission from ©University of British Columbia Press.

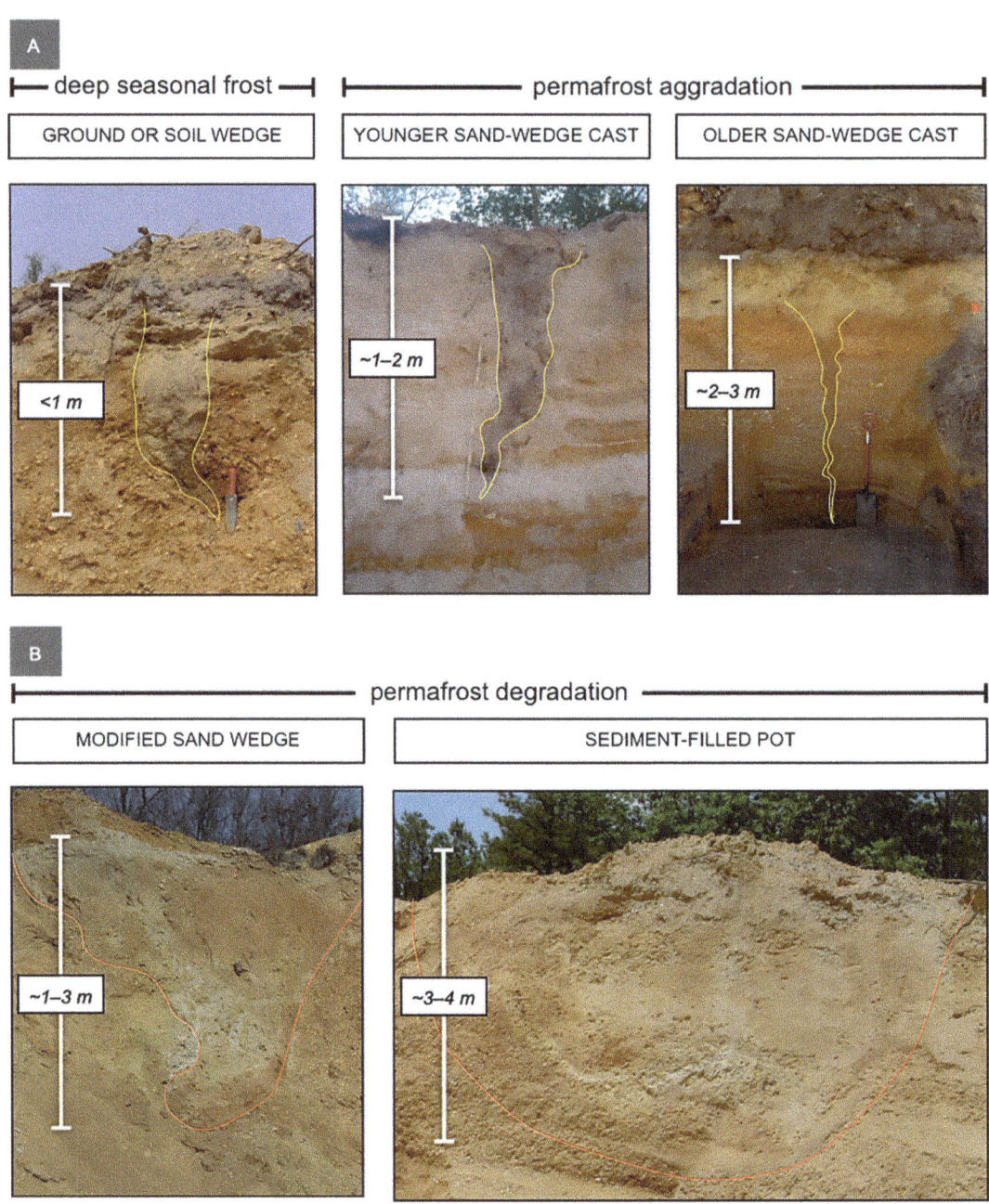

Fig. 2.14 Periglacial macrostuctures and climate oscillation.

(A) A suite of Pinelands periglacial macrostructures that are permafrost aggradation indicators associated with Pleistocene climate cooling (outlined in yellow). Their geomorphic regime is attributed to climate conditions that ranged (left to right) from the advance of deep seasonal frost to discontinuous permafrost to continuous permafrost under cold, dry, and windy conditions (French et al. 2003).

(B) A suite of Pinelands periglacial macrostructures is associated with climate warming and permafrost degradation; where the above aggradational macrostructures—after genesis—have then been modified and deformed by ice-thaw (thermokarst) occurring with climate amelioration (outlined in red). Some are chimeras—complex entities often composed of blended A and B features—that are subsequently further reworked and reformed by polycyclic cooling and warming episodes. Sediment-filled pots are thermokarst bulbs that form at polygon wedge intersections (French et al. 2005). A deformed wedge (above B, right image) propagates off to the right to in-furrow connect another sediment-filled pot that is 10 m away (Figs. 2.3–2.4).

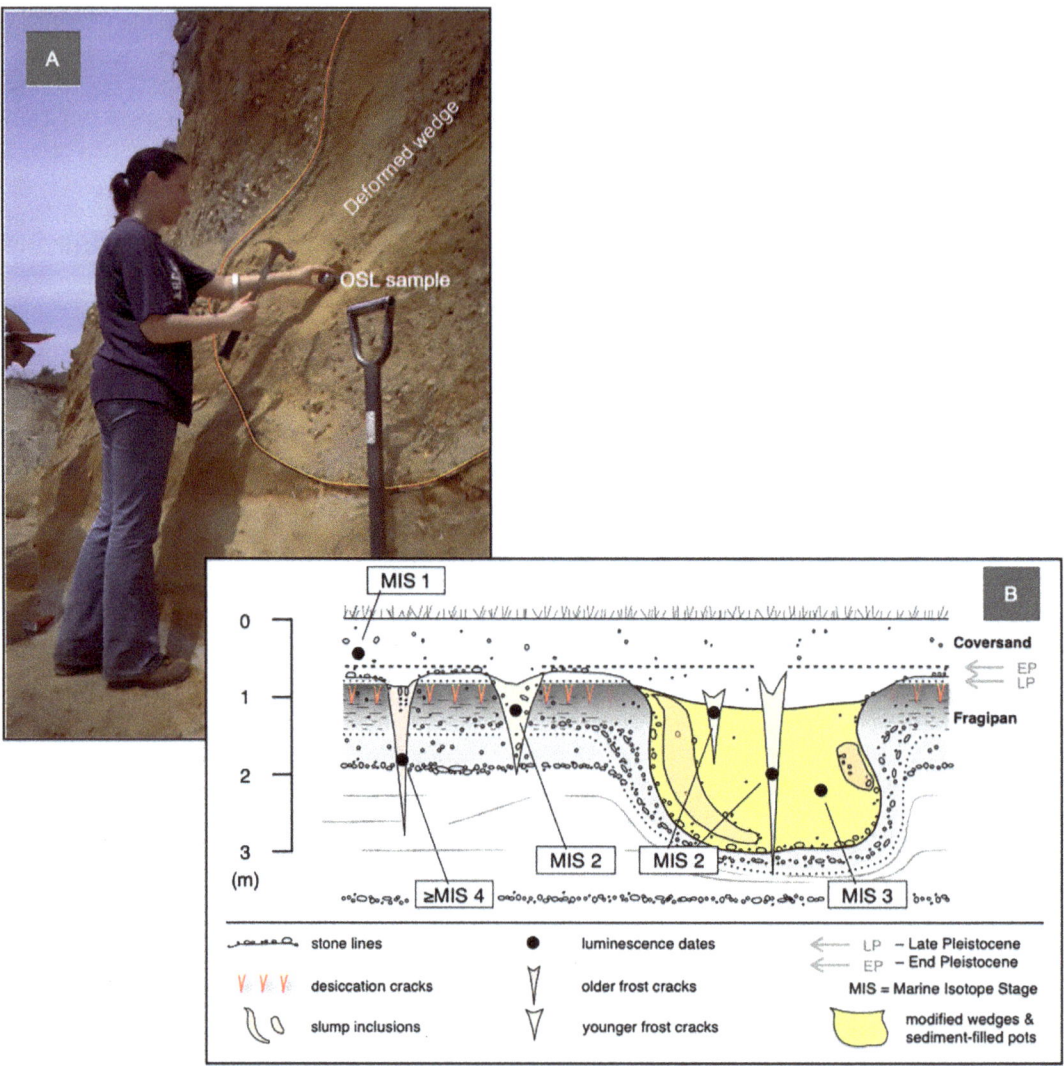

Fig. 2.15 Optical dating of periglacial structures.

In the absence of Pleistocene organic remains, radiocarbon dating in the Pine Barrens has been problematic. In its stead, the optical dating of quartz and feldspar grains has provided an opportunity to time the aggradation and degradation of past permafrost. Silica grains can be chronometers of time passage by the trapping of electrons in grain-surface traps. Like a computer chip (silica), quartz grains (silica) have an affinity for electrons. If briefly exposed to sunlight during wind transport, the quartz grains loose their previously captured electrons. The time clock is reset. Frost cracking and ground thaw allow the burial of grains with reset clocks. Slowly and steadily the sand grains again fill their electron traps through background radiation. The passage of time since grain burial can then be determined in a laboratory.

(A) Collecting a sand sample for optical dating of a deformed wedge (outlined in red), Unexpected pit, Newtonville, NJ. The sampler must be careful not to expose any grains to sunlight or else their clock might unexpectedly reset giving spurious results.
(B) This author, in cooperation with colleagues, has currently collected and processed over 28 OSL samples, with a dozen more pending. A chronology can now be proposed for Late Pleistocene permafrost aggradation and degradation. Figure modified from Demitroff et al. (2012).

Fig. 2.16 Southern extent of Pleistocene past permafrost.

At the turn of the twenty-first century, the prevailing thought was that Pleistocene permafrost—if even had been in extant at all—would be limited to a narrow band ~25–50 km (~15–30 miles) beyond the ice sheet terminal margin, a narrow frozen that laid north of the Pine Barrens (see French 1996: 229, Fig. 13.6).

(A) French et al. (2009) published their southernmost thermal contraction wedge at Fruitland pit, near Salisbury, Maryland. It was a narrow, +1-m deep sand-wedge cast fossilized in dense fragipan. This wedge structure terminated in loose sand in a beard formed as a series of individual sand veins splayed into the loose sand. The formation of these sand veins in such a loose deposit could only have occurred if frozen ground was present during cracking. Fruitland pit is some 260 km (160 miles) south of the Laurentide Ice Sheet's maximum advance.
(B, C) In exploration, this author later conducted an exploratory field trip into eastern North Carolina. A provocative series of wedge-shaped sand-filled structures

appeared to form polygonal patterns upon an upland sand terrace. These wedge structures were hosted within decayed peat, which in turn was itself hosted within fine-to-coarse silty sand. In light of new reporting of potential past permafrost's extent to the Carolinas (French & Millar 2014; Lindgren et al. 2016; Marshall et al. 2021; Merritts & Rahnis 2022), it may be possible to invoke a periglacial component to wedge genesis some 650 km (400 miles) south of the Laurentide Ice Sheet's maximum advance. The Dirt Shop pit, Ocean, near Newport, North Carolina.

Soggy Ground

03
Holocene Hydrogeographic Features as a Resource for Cultural Ecology

In the Pine Barrens of southern New Jersey, the four traditional terms "spungs," "cripples," "blue holes," and "savannahs" are distinct and separate wetland types, important resource sites to early inhabitants, genetic manifestations of periglacial processes, and disappearing as a result of a declining regional water table. Spungs, cripples, blue holes, and savannahs are archaic words, the origins of which are long forgotten. In a modern context, they seem out of place. To the woodcutters who originated and used these terms, they conveyed a specific "sense of place" when they spoke their vernacular. Their economic geography (i.e., how man adjusts to and modifies his environment while making a living) can help us to understand these landforms. Slight changes in the terrain were important to note in the search for forest products, for watering livestock, or finding game (Berger & Sinton 1985; Hufford 1986; Demitroff 2014). A problem arises in attempting to match each wetland feature with an appropriate traditional denotation.

    A review of archival material and local records such as wills, deeds, road returns, historical documents, and maps help to pair place names with landforms and other geographic interpretations (Brown 1943; Norton 1984). Interviews with hunters, foresters, farmers, and other longtime Pinelands residents supplement written material (Myers et al. 2003). This author spent his formative years exploring, fishing, and wading through the Pinelands when not working at the family farm, and spungs, cripples, blue holes, and savannahs were an intimate part of his childhood milieu, which adds local knowledge (Bassett & Zimmerer 2003: 101–102; Bélisle et al. 2018).

Archival and anecdotal materials are used to collect appellatives given to specific wetland locations in the Pine Barrens. Locals often gave critical attention to form and pattern as they foraged for natural resources (Sauer 1956: 290). In names, there are clues about the distinct nature of various wetland features (Sauer 1956: 290-291; Cohen 1983: 15-22). What were the interactions between people and places? Who were the people using these wetland resources? When were they here? Why did they choose to exploit these habitats? What do we know about their qualities, and why is there is such rich lore surrounding them?

Local inhabitants also used legends to give meaning to otherwise unexplainable natural phenomena they encountered. By combining folkloric input (Duffin & Davidson 2011) with archeological and historical data, this facet of a Pinelands traditional way of life can be chronicled (Berger & Sinton 1985; Moonsammy et al. 1987; Myers et al. 2003: 88-90). There is evidence that precontact cultures utilized these features (Bonfiglio & Cresson 1978; 1982; Mounier 2003: 30-33; Cresson et al. 2006; & Mounier 2009). Additionally, the insight of local botanists and herpetologists is also considered, as these features are important habitat for much of the region's rare flora and fauna (Alexander 1951; Laidig et al. 2001; Gordon & Demitroff 2009; Smith 2012; Bunnell et al. 2018).

After matching descriptive monikers with appropriate Pinelands feature, the specific landform's characteristics, morphology, size, position, orientation, and extent are reviewed. With the derivation of their physical attributes, and an accounting of their arrangement in place, the physical features are related to geological and pedological structures. Past studies of Pinelands wetland types were based on specific characteristics of interest to their authors (Harshberger 1916—vegetational; Waksman 1942—peat; McCormick 1979—vegetational; Zampella & Roman 1983—vegetational; Roman et al. 1985—vegetational; Ehrenfeld 1986—vegetational; Collins & Anderson 1994—vegetational; Epstein 1997—hydrological; Laidig et al. 2001—herpetological; Lathrop et al. 2005—herpetological; Palmer 2005—vegetational; Smith 2012—vegetational). None of these studies are particularly suited to geographical pursuits. They were founded on how plants or animals interact with their physical environment, not with man's relationships with the landforms (Barrows 1923: 4-5), save Marsh et al. (2019) on railroad era ethnic settlement.

Work by this author referenced in a geomorphological context is derived in large part from his work with Hugh French (Figs. 3.1-3.2), including French and Demitroff (2001, 2003, 2012), French et al. 2003, 2005, 2007, 2009a, 2009b, which includes: 1) description of paleo-periglacial traces from resource extraction pits—e.g., thermal contraction cracking, thermokarst involutions, structures associated with ironstone deposits, ventifacts); 2) dating of eolian sediments through optically stimulated luminescence; and 3) granulometry, microtextural analysis, and roundness analysis of sand grains. Technical knowledge of landform evolution has been considered "an indispensable prerequisite to successful geographic work" (Barrows 1923: 5).

The apparent recent desiccation of these shallow wetlands is also addressed. Accurate records of shallow aquifer levels in the Pine Barrens until recent times were scanty, episodic, and of a duration too short to document long-term trends in groundwater base level (Epstein 2003). Recent modeling indicates that southern New Jersey's groundwater use is depletive in terms of about 1.3 cubic miles (Konikow 2013). Preliminarily, it is suspected that over-withdrawal from the Kirkwood-Cohansey Aquifer by the Atlantic City region is depleting streamflow and groundwater across watershed boundaries (NJDEP 2003a: 54). In the absence of long-term groundwater data, an account of the physical condition (e.g., depth of water-fill; duration of water-fill; complete absence) of spungs, cripples, blue holes and savannahs can be used to suggest historical changes in groundwater amount whether due to climate change (Brooks 2004; Graham 2016) or over-withdrawal of groundwater (French & Demitroff 2001). The shallow wetlands are considered "windows" into the Pine Barrens water table, since they are quite sensitive to groundwater changes (Brooks 2004; Graham 2016; Calhoun et al. 2017).

By studying the data collected for the cultural aspects of these features (i.e., their etymology and their usage), certain inferences can be deduced about the wetlands past physical condition. Were they once more prominent in the landscape? When did changes occur in them? Are they cyclic in nature? What proportion of the change is anthropogenic?

Fig. 3.1 Pinelands geodiversity is critical to Pinelands biodiversity.

Pulitzer-prize winning author John McPhee (1968: 56) famously said this about traditional Pine Barrens residents, "they know that their environment is unusual, and they know why they value it." It is an environment rich in biodiversity, which makes for a vibrant cultural landscape too. This outstanding biodiversity is predicated upon an equally outstanding geodiversity (Gray 2013, the valuing and conserving of abiotic—non-living—nature). A wall of sand-wedge casts on display during a field excursion for the Twentieth Annual Meeting of the Geological Association of New Jersey, October 10–11, 2003, hosted by Richard Stockton College of NJ (French & Demitroff 2003: 131–137. Photo courtesy Julie Akers.

Fig. 3.2 Life on the edge.

Ice-marginal conditions—the consequences of being next to a continental ice sheet during an ice age, remain poorly understood. During the Pleistocene, which lasted about the last two-and-a-half million years, about 51 cold episodes came and went (Head 2019). In New Jersey, the last of the ice sheet advances reached it maximum advance in New Jersey about 25 ka and began its retreat by 24 ka (Stanford et al. 2021).

The Laurentide Ice Sheet's terminal moraine is marked by a 67 m (220 feet) hill topped by a water tank between Exits 10 and 11 on the New Jersey Turnpike. The photo was taken by the author ~1982—before sound barriers were built that now partially obscure the view.

04
Spungs of the Pine Barrens

Etymology of Spung

The earliest mention of utility to the closed-depression wetlands of South Jersey (Fig. 4.1) as a distinct landform type for cranberry culture was in Barber and Howe (1868: 531, winter ponds, or "slashes"), and in a monograph by White (1870). The latter work initiated a speculative land rush (cranberry fever) to develop bogs in the Pinelands (Eck 1990: 7–11). In a "newly enlarged edition" (White 1885: 29–30), White suggested that these "heath ponds" or "low basins" are opportune locations for berry cultivation. They were described as winter-flooded pools underlain by a "hardpan."[1] Later soil surveys (Engle et al. 1921: 41; Burke et al. 1929: 1699) mentioned the basins as agriculturally unproductive places where only a few people have made the effort to grow cranberries.

Use of the word *spung* appears to be confined so far to two locations, southern New Jersey, particularly the Pine Barrens (Neuendorf et al. 2005: 621; Hall 2012: 219), and to the sandy Barrens-like terrain of Massachusetts.[2] Although the pronunciation is widely accepted to

---

1. Hardpan is a general, non-technical term for "a relatively hard, impervious, often clayey layer of soil lying at or just below the surface, produced as a result of cementation of soil particles by precipitation of relatively insoluble materials . . . " (Neuendorf et al. 2005: 292). Spung bottoms are not cemented but are hard and fragipan-like.

2. From Groton, Massachusetts, natural historian Samuel Abbott Green reports *spungs* (1879: 8, 39, 44—aka *spongs* (Green 1879: 8, 50) or *spangs* (Green 1879: 8, 29, 30, 32; 1894: 227; 1912: 57), in association with local wetlands. He deems the term obsolete, citing *Suffolk Words and Phrases* by Moor (1823), and meant "an irregular, narrow, projecting part of a field, whether planted or in grass." This suggests an English derivation for the

rhyme with the word "rung" (Radis 1987: 19; McPhee 1968: 61, "sung"), its written form is problematic. Maps and deeds often used "spung" during the seventeenth and eighteenth centuries,[3] but by the nineteenth and twentieth century, "spong" was often preferred. The change in spelling may have been spurred by introduction of the "spong" pocket-type meat mincers (later coffee mills) developed by James Osborne Spong starting in 1856 (*Grace's Guide* 2018).[4]

The *Oxford English Dictionary* (OED 1915 & 1914, respectively) listed both spellings. A "spung" is defined as a purse or fob, a small "pocket," formerly made in the waistband of breeches, for carrying a watch, money, or other valuables (Fig. 4.2). It featured a composition of wash-leather or stout lining material (Cunnington 1972: 69) and was sewn in just under the waist (personal communication, Jenny Lister, Museum of London, August 8–12, 2003). The verb "to spung" someone means to pickpocket them (OED 1915). Breeches were popular menswear from the sixteenth through the early nineteenth centuries, worn by men of all socioeconomic levels (Colonial Williamsburg Foundation 2018). Illustrations of a spung pocket are difficult to find for two reasons. First, the coat and the waistcoat cover this portion of the male figure. Displaying this portion of the garment would have involved a very informal pose not in keeping with social conventions of the time. Second, by the latter half of the eighteenth-century breeches had come to be called "small clothes" and were not overtly displayed. The coat, the waistcoat, and the cravat (neck cloth) took precedence (personal communication, Kathleen Gossman, Furman University, August 1 & 18, 2003; Colonial Williamsburg Foundation (2018), an eighteenth "century man was almost never seen without his waistcoat").

In contrast, the *Oxford English Dictionary* (OED 1914) defines a "spong" as a long narrow piece or strip of land,[5] which does not conform to usage in South Jersey. Spong is also an obsolete word for a drift-way (drove

---

term 'spung.'

3. Early Swede and Finn settlers used the term panner (*Svenska Akademiens Ordbok* 1952: Vol. 19, 127, a low, circular-bottomed boiling pot) to describe "broad and deep-water basins," apparently spung-like ponds in salt meadows (Craig & Williams 2007: 68).

4. Spong coffee mills are occasional encountered in the Pines. The author's spong mill has been in use since 1978.

5. Moor (1823: 387) adds spong in association with queech—untillable parts of a field that lack trees or undergrowth, or squeech—untillable parts of a field with underbrush; not unlike arable land with spungs upon it.

road) for cattle (OED 1914). Environmental consultant McCormick (1979: 231) defined spungs as "nearly circular depressions that are scattered through the region." This author agrees with both McCormick's spelling and description, which McCormick likely learned from his mentor, Silas Little, the eminent Pinelands forester (personal communication, Ted Gordon, 2002). In describing the cultural geography of the southern Pine Barrens, Marsh (1979: 192; et al. 2019) used "spung" to describe presumed "cave-ins" at stream heads. Sneddon et al. (1995: A-30) called enclosed basins in the Pine Barrens "spungs."

Pinelands spungs are simply small-enclosed "pockets," often accommodating water for at least part of the year. In Ocean County, the term "pocket bog" is often used to describe a spung. A grizzled old denizen of the Pines reflected that the term might refer to the ease of preparation that pocket bogs afforded for cranberry culture. As spungs are often bare of woody vegetation, one could sprig out a planting (Everett 1981–1982: 3208, propagate with small pieces of stolons with roots and leafy shoots) with minimal investment and without the backbreaking toil of turfing. Consequently, the cultivator could pocket the savings in labor (personal communication, David Emory, 2002; also see White 1885: 84–85, 96–100). However, this author believes "pocket bog" is a direct translation of the now obscure word "spung." Other monikers include: bogs (Alexander 1951); prim ponds = privet ponds, for the privet-like Leatherleaf (*Chamaedaphne calyculata* [L.]) growing there (OED, March 2007); goose ponds = larger spungs (Beers 1872a); duck ponds = smaller spungs (Townsend et al. 1873); elephant wallow or hog wallows; whale wallows (Radis 1987: 19); hog holes (Hartman 1950, 1978: Map 4), holes; winter ponds (Burrough et al. 1793); watering places (Harshberger 1916); heath ponds (White 1885; Waksman et al. 1943); dry ponds (McCormick 1979); wood ponds (Smith 1773); savannah ponds (Paul 1822); cedar ponds (Clement 1852: 69); and fens (Collins & Anderson 1994).

Apellatives of Spungs

Many of the individual ponds had descriptive names, most of which are long forgotten. Ancient deeds, surveys, and maps often employed these features as important markers designating property boundaries. Old road returns are excellent resources, since early byways were often modified Indian trails, on their way from pond to pond. Gunning clubs are also good sources for uncovering the names of spungs, as they remain an important resource for hunters to this day. Older local residents remember the names of childhood fishing, skating, and swimming holes, now consigned to near-oblivion.

Hunting themes are common to the pools. The Bears Hole—Pancoast Mill (Fowler & Lummis 1880), Crane Pond—Mays Landing (Wright? c.1867; Gifford 1896: 162), Big Goose Pond—Egg Harbor City (Wherry 1957) (Fig. 4.8C), and the Hog Holes—Millville (Hartman 1950, 1978: Map 4) are all obviously named for the game that could be stalked there. Others are more archaic, like Sound Pond—Hammonton (Beers 1872b), an obsolete hunting term for a spring or pool of water (OED 1913). Spungs can also have geographically descriptive names, such as Desolation Pond—Nesco (Clement 1845: 58), Round Pond—Milmay (Fowler & Lummis 1880) (Fig. 4.5A), and Punch Bowl[6]—Milmay (Demitroff 2003). Blue Bent Pond—Mizpah (Thatcher 1919) was a perfectly circular pond. Bent[7] is derived from the curved portion of a rounded form, now rare in usage (OED 1887). Bears Head—near Mays Landing (Harshberger 1916: 143) is reputedly named for an unfortunate encounter with the last of South Jersey's bears (Bailey 1999: 12), but instead receives its name from a large pond at the head (source, beginning, or upper part of a stream) at the uppermost reaches of a Bear Branch.[8]

Another theme relates to the importance of spungs as way-stops for thirsty travelers. Hospitality Ponds[9]—Williamstown/Squankum (Clement

---

6. In Massachusetts Green (1912: 64) used "Punch Bowl" to describe both the depression itself and its locality in the manner that Milmay residents do in the Pine Barrens north of Milmay, New Jersey.

7. Along England's Suffolk coast, within the European Sand Belt, are "bentles"—flat sandy lands that only support the growth of coarse unprofitable bent or couch-grass (e.g., *Elytrigia juncea* [L.] Nevski, sand couch-grass), reedy sedge-like grasses that even hungry cattle reject as pasturage (Moor 1823: 25–25), offering an alternative pond namesake for poor pasturage due to the presence of coarse grass or sedge.

8. There were at least three Bears Head ponds and Bears Head Branches along a trail by the same name (today's Bears Head Road), one at Cornucopia Avenue (Fig. 4.8A), one at Doughtys Tavern, and one at Baker Road. Likewise, the head pond on Calf Branch (Milmay) is the Calfs Head. A "head" is the source or headwaters of a stream. To "head" a stream or water body is to go around it (OED 2013). Bear also relates to wood so poor that is not worthy of a day's labor (Moor 1823: 505), cf. Littleworth for Petersburg (Gordon 1850).

9. Clement (1888: 186) writes, "These ponds were at that time about the head-waters of a stream by the same name and covered considerable territory. Sometimes they were made by the beaver, but generally lay upon the low, flat soils peculiar to land in South Jersey." The pond-complex property was "beside the Old Cape [Tuckahoe] Road."

1888; Stevenson 1890: 38), Horse Break Pond—Buena (Wright 1867a) (Fig. 4.3), Jacob's Well—Williamstown (Gloucester County Road Return c. 1790), Watering Place Pond—Warren Grove (Harshberger 1916: 143–144) (Fig. 6.9), Swains Stopping-Off Place—Hunters Mill (Hartman 1978: Map 8) (Fig. 4.14A), and the Oasis—New Italy, with another at Tar Kiln Neck, are good examples of resting places. Others are not so obvious, like Wall Pond at the head of Tar Kiln Branch near Belleplain (Cook & Vermeule 1890: Map 15; Harshberger 1916: 143), which may relate to an obsolete nautical term. The pinewoods surrounding this pond were kilned for naval stores production (Folger 1989: 71), essential materials for the local shipyards and exportation to England. To "lie at the wall" is to rest at dock (OED 1921). Wall Pond likely served as a resting spot for tar kilners like nearby Tar Kiln Johney (Hartman 1978: Map 1, lived there in 1778). Keeping with the nautical theme, there was a Sailor Boy Tavern by Tar Kiln Neck, between Hammonton and Elwood, NJ (Anonymous 1880: 3).

Few places in the Pines could provide access to potable water as good as that of a spung. Vermeule (1894: 277) noted, "It must be remembered that in this sandy soil a very slight depth of water is usually sufficient to destroy all vegetable growth, and shallow ponds keep as sweet and pure as the deeper ones." The small, underfit (misfit) streams (Dury 1965) in this region are located in broad swamps adjacent to modern waterways. These wide swamps have been interpreted as paleochannels (Farrell et al. 1985; Newell & Wyckoff 1992; French & Demitroff 2003, 2012; French et al. 2005; Demitroff et al. 2019) that relate back to water flowing over frozen ground during the Late Pleistocene. Getting your livestock to a stream also had its perils. Animals had to negotiate the wide swamp (paleochannels) to reach the open water of underfit stream channels. Main Road in Vineland was previously known as Horse Bridge Road, in recognition of several horses that permanently mired while trying to cross one such broad floodplain on their way to the Blackwater Branch, sometime prior to 1781 (Hampton 1917; Ackley 1945).

Spung Morphology

Spungs vary in form (Figs. 4.4–4.9). Many spungs are rounded in plan view, while others are irregular. A few display classical oriented forms reminiscent of Carolina Bays (Moore et al. 2016). Round forms are interpreted as nascent spung shapes, which can coalesce into a kidney-bean-shaped or "reniform" outline, like at Winks Meadow (Fig. 4.7) by the Mankiller Swamp. It is hypothesized that drying katabatic winds were dominantly NNW–SSE year-round, without a summer switch to warmer, moisture bearing Westerlies (see Markewich et al. 2015: 167, Fig. 22c, July paleowind direction during the Late Glacial Maximum). In evidence, the pavement einkante ventifacts as named and documented in Demitroff (2016) only form as an erosional end product of unidirectional wind (Vârkonyi et al. 2016: 29). In consequence, during cold dry periods spungs have a weak wet or "pluvial" stage. It is both the absence of windblown summer water currents (lack of hydrofill) and an armoring provided by spung-encircling fragipan that forestalls the basin orientation observed in the Carolinas (Kaczorowski 1977) or in northern Alaska and Canada (Black & Barksdale 1949; Carson & Hussey 1962; Côte & Burn 2002), although oriented lakes too may initiate during non-pluvial dry cold periods (Wolfe, S.A. et al. 2020) The average South Jersey depression covers less than an acre, with the largest reaching a hundred acres or more (e.g., Broad Pond—near Elmer,[10] Fig. 4.6D; Lookout Pond—near Mays Landing, Figs. 4.9C, 4.10). See Mihalasky and Del Sontro (2003) for examples of spung-area measurements. Like northeastern China (Zhang et al. 2007a; 2007b), Pine Barrens deflation basins occur at different scales, vary in type, and appear to be at varying stages of development. Cold-region deflation basins—blowouts—can even occur in the presence of coarse gravel in areas devoid of plant life (Seppälä 1971: 25)

Spungs periodically flank the terraces of many Pinelands streams, rivers, and nearby uplands (Figs. 4.11–4.12). A few are riverine, and those tend to occupy the uppermost channels of smaller streams. On the higher positions between rivers (interfluves), many small and medium-sized spungs have parabolic bottoms, becoming flatter as depression size increases. Spungs that occupy low positions in the landscape, such as those in broad paleochannels, have remarkably planar floors. It is possible to explain this spung basin morphology by a permafrost-eolian model. When permafrost conditions existed, higher terrain on the interfluves probably had low ice content, which poorly bonded the enclosing sediments, and the lack of a strong ice bond allowed strong winds to create a classic parabolic deflation basin shape as can be seen in active coastal dunes today (Jungerius & van der Meulen 1989: 370, Fig. 1). Rhodehamel (1970: 10) estimated that approximately two percent of the Pine Barrens

---

10. A much larger spung-like water body is depicted between Pittston (Elmer) and Pole Tavern that remains enigmatic (Watson 1812; Burr & Illman 1835) forming the head of both the Salem River and Palatine Branch.

surface is covered by enclosed basins. He interpreted many of these basins as deflation basins (personal communication, Edward Rhodehamel, February 6, 2007). Zhang et al. (2007a: 69) found grassland deflation basins in Hulunbuir (northeast China) can initiate along the line of sand wedges (Zhang et al. 2007b, sandy soil wedges [perhaps ground wedges?[11]]).

Early Spung Use and Topography

The association of spungs and their use by prehistoric peoples is well established (Bonfiglio & Cresson 1982; Mounier 2003; Cresson et al. 2006), especially as it relates to the depressions found on adjacent coastal plain outside the Pine Barrens (Figs. 4.12–4.13). Outer Coastal Plain sites are likely underrepresented. Construction occurs preferentially on the heavier sediments of the Inner Coastal Plain, outside the Pinelands National Reserve's border. Disturbance by construction at such sites often necessitates environmental survey work, providing archeologists numerous opportunities to investigate the many depressions like those described by Bonfiglio, Cresson, and Mounier (Tomaso et al. 2008). The spungs of the Pine Barrens have however been passed over. The historical absence of development pressure corresponds to a lack of economic impetus for their scrutiny (personal communication, Jack Cresson, 2002).

At slight variance with the observations of Bonfiglio & Cresson (1982: 35, Fig. 6B), the Pinelands spungs tend to be slightly "rimmed" or raised towards the southeast, and do not share the southern and western bias of rim use noted on Inner Coastal Plain depressions. It is the author's impression that use of Pinelands watering holes favored the south to southeast positions, this related to a more modest topographic rise than their Outer Coastal Plain counterparts. Farmer Ron Bertonazzi of New Italy section of Vineland, New Jersey, has collected an assortment of precontact artifacts relating to the Archaic (c. 12–3 ka) and Paleoindian (14–13 ka) Periods (Cresson 2000) from a spung complex known as Vannaman's Thick & Hole Tract (Hartman 1978: Maps 4 & 7).

A marked association exists between Indigenous trails and spungs in the Pinelands. Paths linked these watering holes in a chain of way-stops for the earliest inhabitants (Fig. 4.14) along "old Jersey" roads.[12]

---

11. Ground wedges are associated with deep seasonal frost and not permafrost, Wolfe et al. 2018)

12. Old Jersey roads were described as "miserable sandy roads . . . anything but . . .

For example, numerous such connections are apparent in Buena Vista Township, Atlantic County. The Hance Bridge trail (Hartman 1978: Maps 4 & 7), within a distance of 6 km (4 miles), linked the Oasis spung in Vannaman's Thick & Hole Tract—New Italy, with the Lummis Swamp spungs—Richland (Denny 1774, French & Demitroff 2001: 340, Fig. 2B) (Fig. 4.8B), and with an as yet unnamed cluster of spungs north of Richland, at St. Augustine's Preparatory School. From here, a hunter could continue to the Bears Hole—Pancoast Mill (Fowler & Lummis 1880). Alternatively, the traveler in Richland could take the Cohansey trail (Purvis 1917: 27–29; Hampton 1917: 49–50; Stewart 1932: 91), which intersects at the unnamed spungs above, westward 1.5 km (1 mile) to Horse Break Pond—New Rome section of Richland (Wright 1867a; French & Demitroff 2003: 118, Fig. 2) (Fig. 4.3), another mile to the spung at the 1767 Parson plantation (personal communications, Glenn Bingham, Mark Parsons, August 06, 2018; also see Parsons 1767[13]), later Campbell's Tavern—Buena (Boyer 1962: 155; Boucher 1963: 39), and then an additional 6 km (4 miles) to Egg Harbor Pond—Vineland (Hackney 1795, French & Demitroff 2001: 340, Fig. 2A).

In England, shallow round watering holes (dew ponds) with prehistoric usage (e.g., Cissbury Ring, Chanctonbury Ring, Stonehenge) were connected by ancient trails (Hubbard & Hubbard 1907; Martin c. 1915; Toms 1927; 1934a; 1934b). Differences of opinion existed about whether dew ponds were natural or anthropogenic, and about their antiquity (Allcroft 1908, 1924; Johnson 1908; Beckett 1909; Pugsley 1939; Hayfield & Brough 1986). All agreed that basin (re)construction continued into modern times (Herschy 2012), as they served as important watering holes in the high and dry chalk grasslands of the South Downs (Landscape Design Associates for the Countryside Commission 1996: 21). Most dew ponds have recently dried up (Wood et al. 2003; Grimwood et al. 2004).

The South Downs had a periglacial environment during much of the Late Pleistocene, sharing many cold climate features (Catt 1987—frost cracks; Williams 1987—dry valleys; Murton & Lautridou 2003—involutions; Murton & Ballantyne 2017—past permafrost) with South Jersey. Depres-

---

enjoyable" (Stewart 1918: 72). In example, into the 1860s the Cohansey trail remained an important byway, yet Warner (1869: 17; reprint: 20) recounts, "at every step the rider was regaled by the musical click of the grub oaks rattling against the carriage wheels, or having his nosed tickled by the sharp, stiff feathers of the pine trees."

13. Parsons also lived at Parsonstown by the Parsons Ponds (Figs. 2.5C, 4.8A). This property was 11 km (7.5 miles) south of the Parsons Plantation along the Blue Anchor trail. Midway between Parsonstown and the Parsons Plantation, relations (Vannamans) kept livery at the New Italy Oasis in the Thick & Hole Tract.

sions described as periglacial basins have been reported in southern England (Sparks et al. 1972; Hutchinson 1980; Boreham 1996, Boreham & Horne 1999; Boreham et al. 2010; Toms et al. 2014), although it is difficult to separate man-made or man-modified basins from natural ones (Prince 1961; 1962; 1964; 1979; Williams 1973; French 2018: 335–336). Other European periglacial basins have early cultural associations during the Neolithic (Collin & Godard 1962) and Roman (Steinhausen 1936) periods. There is a modern relationship between such basins and rural roads in Belgium (Gillijins et al. 2005).

In Delaware, adjacent to South Jersey, enclosed depressions known as Delmarva Bays (Fenstermacher et al. 2014; 2015) are considered analogs of Pine Barrens spungs (French et al. 2005: 10). Of Delaware's basins, Rasmussen (1958: 18) stated,

> Early settlers built their cabins on the better-drained sandy rims, and, in colonial times, mansions were built at the higher rises where rims coalesced. Early roads followed the sandy rims in arcs, and early cultivation extended the fields along the rims and encroached upon the basin centers in which some drainage could be established.

Bay/basin features have evidence of precontact use in Delaware (DEL-DOT 1993).

The relationship between spungs and cranberry bogs was introduced earlier in **Etymology of Spung**. Continuing, the author of *Cranberry Culture*, White (1870, 1885), owned a large tract of clustered spung swampland where wild cranberries grew in depressional patches (spungs) known as Rake Pond Tract (Fig. 4.15). In 1860, Joseph White planted his first vines in this location, beginning a lifelong successful career in cranberry culture. Joseph met his future wife, Mary Fenwick ("a comely virgin"), while scalping turf at a spung site (White 2014). Cranberry culture began in duck and goose ponds—two of many terms used to describe smaller and larger spungs, respectively—before the industry moved to sluice-gated bank farms sited upon Pleistocene braided paleochannels and alluvial fans. Surrounding dunes were exploited for the sanding of cranberry vines, which rejuvenates old plantings.[14] Here sand quality is important. It must be free of silt and clay (Eck 1990: 170), which by nature dunes are

---

14. Sanding also helps warm a bog providing frost protection; as well, the increase in sunlight reflection corresponds to an increase in photosynthesis (Eck 1990: 224).

due to wind sorting. Carolina Bays, too, experienced agricultural exploitation—e.g., Lake Phelps, c. 1784 rice production (Richardson 1981: 25).

Origin of Spungs

The enclosed basins were first scientifically scrutinized on a Friends of the Pleistocene field trip (Jones 1949: 4; Peltier 1949a: 6).[15] Wolfe later suggested two hypotheses for the origin of these basins, many of which surrounded "Wolfe's Farm"—a family fruit operation—just east of Hammonton in Nesco (Olsson 2001). Wolfe (1952—reissued 1954, 1953) asserted that many of these internally drained shallow depressions were present throughout the New Jersey Coastal Plain. He first attributed them to sediment subsidence because of freezing and thawing in a periglacial climate during the Pleistocene. Wind action, mass wastage, and erosion further modified them after their formation. Wolfe (1956, 1977) later reinterpreted them as thermokarst lakes or thermokarst basins in response to critics. The widespread occurrence of loess, sand dunes, and ventifacts further indicate that New Jersey was "a treeless tundra plain with strong, sweeping wind action during the Pleistocene Epoch" (Wolfe 1977: 293). Some continued to interpret these depressions as thermokarst basins (e.g., Stanford 2005; 2017, "melting" of permafrost).

In response, Rasmussen[16] (1953), considering a periglacial hypothesis to be too speculative, argued that the basins in New Jersey were similar to the more southerly "Carolina Bays[17]" and "Delmarva Bays." To Rasmussen, it was highly unlikely that these southern bays could share a periglacial genesis with Wolfe's depressions because of their distance from the continental ice sheets.

Numerous theories have been presented in over 350 publications to explain the origins of the Carolina Bays (Ross 2000). Early on, Rasmussen (1958: 95) reported that a Delmarva Bay in the Bothwick (north-central

---

15. Peltier 1949b concurrently reported large 0.6 X 1.5 m (2 X 5 feet) wedge structures in Pleistocene sediments on a terrace of the Susquehanna River in Pennsylvania, indicating that interest in past permafrost was waxing at the time of the Friends of the Pleistocene trip.

16. William "Wild Bill" Charles Rasmussen was killed by a land mine in South Vietnam on May 11, 1973.

17. Otvos (1973) reports Carolina Bays (Citronelle Ponds) in association with Pleistocene parabolic dunes (Otvos 1995, those dunes continuing to Louisiana) in the Florida Panhandle, and attributes their genesis to wind-action.

Delaware) area appeared to have a frost cracked and involuted soil horizon, perhaps having been affected by frozen ground during the Pleistocene. However, he again concluded that a periglacial hypothesis was untenable for the origin of the Delmarva Bays because they extend so far south (Rasmussen 1958: 165, 170). A little later Brunnschweiler (1962) speculated that southern bays were indeed associated with frozen ground, as has Swezey (2018).

In his inventory for the National Park Service, McCormick (1970: 83) recognized the perplexing nature of the Pinelands basins:

> [The] numerous saucer-shaped depressions scattered throughout the Pine Barrens. These depressions, similar in many respects to the better known Carolina Bays, have been attributed to phenomena of original deposition of the Coastal Plain formations, to differential illuviation of clay and silt particles and their deposition in the depression areas, to wind erosion, and to frost-thaw action during the Pleistocene.

McCormick (1979: 231) was the first author to call these "nearly circular depressions that are scattered throughout the region" after their local appellation, "spungs."

Two archeologists who studied the relationship between the closed depressions and the settlement patterns of various precontact cultures (Bonfiglio & Cresson 1978; 1982) revisited Wolfe's thermokarst-lake-basin hypothesis. A new interpretation of basin origin was put forward. To better characterize these sites, Bonfiglio and Cresson chose a "pingo-scar" hypothesis to explain basin formation. Pingos are one of several forms of frost mounds known to occur under cold-climate conditions (Mackay 1979; Nelson et al. 1992; Ballantyne 2018: 97–115; French 2018: 159–167). The southern rims of the South Jersey basins appeared to be subtly ramparted, which Bonfiglio and Cresson hypothesized was diagnostic of an ancient pingo (also see Russell & Krug 1980: 146). Upon thaw and degradation, pingos are reported to leave depressions rimmed by material sloughed off from their sides (Ballantyne 2018: 113–115; French 2018: 333–335). A "pingo-scar" hypothesis was advanced for many spung-like basins identified in the northern United States (Flemal 1972; Marsh 1987; Stone & Ashley 1992; Stanford 1997; Schwans et al. 2015) and especially across northern Europe (Maarleveld & van den Toorn 1955; Dylik 1964; Pissart 2003; Makkaveyev et al. 2015).

However, it is unlikely that the enclosed wetlands of southern New Jersey are the result of pingo degradation (French & Demitroff 2001). The site-specific conditions necessary for the growth of pingos are lacking, and the number of such features in the Pinelands vastly exceeds that of pingos in Arctic locals, where the site-specific conditions are most advantageous for true pingo propagation.

Custer (1986) credited Bonfiglio and Cresson for recognizing the basins as cultural resources but felt it unacceptable to invoke a periglacial hypothesis for the basins' origin, at least with the information available at that time. He argued that the New Jersey bogs and their counterpart, Delmarva Bays, lacked pollen associated with periglacial landscapes. Moreover, the various models for estimating mean annual air temperature near the ice margins[18] ranged from 2–7 or 8° C, temperatures far too warm for even discontinuous permafrost to have occurred. Finally, Custer believed that there had been considerable changes to the basins during the Holocene, including modification by active groundwater recharge.

Since Wolfe's (1953) inaugural paper, understanding of both Pleistocene and modern periglacial environments has increased substantially. Given advances in the understanding of geomorphological processes, coupled with Wolfe's limited Arctic experience, his two papers (1953, 1956) were—from a modern perspective—ambiguous and vague. Péwé (1973: 22), Washburn (1979: 303–304) and Black (1983: 87) were unconvinced that true ice-wedge casts were ever correctly identified in New Jersey, imparting further doubt about Wolfe's periglacial explanation for the New Jersey basins. The thermokarst-lake-basin hypothesis has been questioned on grounds of the absence of evidence for ice-rich silty sediments in the basins typical of thermokarst terrain or underground drainage associated with thaw lakes (French & Demitroff 2001: 345).

In August 2000, initiated by a December 1999 phone call, a field trip was held to investigate the nature of the southern New Jersey depressions that began the work of French and Demitroff. Cresson (2000) briefly described the excursion. A powdery grey dust was found in the backhoe trench dug across the bottom of one of the Lee Ponds,[19] a small circular spung on Loders Branch of the South River, ~3 km (2 miles) SE of Richland

---

18. See "Palynological Evidence" in Demitroff (2016: 127).

19. The John Lee Place was an eighteenth-century cabin at the intersection of the Cumberland, South River, and Cape trails at a series of spungs (Fig. 4.5B). To much surprise, in 1879 surveyors happened upon 76-year-old "Old Aunt Nancy Lee" living there alone in self-sufficiency "far remote from the abodes of civilization" (Taylor 1879).

(Fowler & Lummis 1880; Hackney 1795—as Green Branch) (Fig. 4.5B). A centripetal drain (Wolfe 1953: 134; 1956: 508) or "bellybutton [name given by this author]" was examined near the center of the spung. The apparent recent desiccation of the Lee Ponds and other wetland features in the region was noted by Cresson, as well as by two newspaper reporters (Kaskey 2000, Hajna 2000a). Cresson then added the observations regarding trail linkage and place-name correlation between the various periglacial features "in a pattern of transhumance that began thousands of years ago" (Cresson 2000: 4).

French and Demitroff (2001: 337) argued that the enclosed wetlands could best be explained by the periglacial wind-action hypothesis (Figs. 4.16–4.17). Spungs appear to be deflation hollows—blowouts—created when strong katabatic winds flowed southwards from the continental ice margin across the sparsely vegetated, tundra terrain of the Pine Barrens. The ramparts of Bonfiglio and Cresson (1978; 1982) were reinterpreted as dunes developed from northerly, then easterly winds (Fig. 4.18). Early cultures used these ponds as camp areas in response to the regional groundwater table returning in response to sea level rise due to the melting of polar ice sheets and a warming (i.e., wetter) climate some 14,000 years ago. Wetlands appear to have dried up as the regional water table fell in response to increased water usage from agriculture and urbanization, and an increasing evapotranspiration rate with climate amelioration among other factors. The spungs were later considered to be similar in age and origin to the Delmarva Bays (French et al. 2005: 10). The widespread presence of Pine Barrens ventifacts and attendant polar-desert features supports a deflational or "blowout" origin to various mid-Atlantic closed depressions (Demitroff 2016: 131; Wolfe et al. 2023).

Primary infill thermal-contraction wedges in the Pine Barrens (French et al. 2003: 260, 261) and beyond to Delaware, Maryland (French et al. 2007; Losco et al. 2010) and Pennsylvania (Merritts & Rahnis 2022; et al. in review) are consistent with the deflationary hypothesis of spung origin. Mihalasky and Del Sontro (2003) assembled, scanned, and associated the physical locations (georeferenced) 21 aerial photomosaics dated c. 1931[20] (Del Sontro 2003). Forty named wetland basins of various

---

20. The c. 1931 (1930–1931?) aerial photomosaics are taken at a time of great deforestation, providing better views of the ground beneath sparser vegetation. Cottrell (1937: 10) states, "The devastation wrought by man's imprudence has vastly changed the character of the forest growth . . . the cedars are merely matchsticks . . . replaced by worthless brush . . . pines are weak and spindling . . . oaks . . . rarely attain sawlog size."

morphologies were delineated and analyzed in preparation for a larger study that would include spung size and geographic distribution. Many spungs disappeared or dried up during the period between the dates of Del Sontro's photography (c. 1931 & 1997).

Narrow channels can link together chains of small round spungs along some stream corridors in the Pine Barrens. They appear as though they were once beaded channels (van Everdingen 1998) since these channels are sharply defined and are generally straight with occasional angular bends (Hopkins et al. 1955: 141). These unusual drainage networks appear to follow remnants of wedge furrows and sediment-filled pots that became intermittent gullies after a period of deflation and bottom erosion (Figs. 2.5, 4.19). Spungs can appear like beads on a necklace, following the traces of sand-buried paleochannels, their loose windblown channel infill facilitates the entrainment of sand by wind. Whether or not these channels were originally associated with ice-wedge polygons, composite wedges or deformed sand wedges has not fully been determined. The author has observed numerous pit exposures where networks of relict furrows and sediment-filled pots crisscross the Pinelands uplands, although Gao (2014: 145) had trouble discerning wedge polygons in Pine Barrens aerial imagery yet noted well-developed wedge-polygon networks elsewhere from southern New Jersey continuing south to the Delaware/Maryland border of the Delmarva Peninsula. Their apparent absence to Gao's view is likely due to the presence of dunes and coversands that obscured their traces, along with the Pine Barrens' dense vegetation. Relict wedge deformation structures (e.g., sediment-filled pots) are most often preserved if dense fragipan aided in their "fossilization" during thermokarst processes (French et al. 2005). This pond linkage between sediment-filled pot-like structures occurs in adjacent states too (see Figs. 2.3, 2.5).

Similar oval-to-linear drainage channels in the semiarid steppe environment of East Siberia are complex and cyclic in nature, experiencing intermittent run-off, eolian activity, cryoturbation, and solifluction (Lyubtsova 1998). Along the coastal plain of Svalbard, Repelewska-Pekalowa (1991; 1996) concluded that piping and thermoerosion produced comparable beaded channels (also see Godin et al. 2014). The hard carapace (Fig. 2.6A, in red) associated with modified wedges and the dense nature of the enclosing fragipan-like band would adequately armor the macrostructures' traces (Demitroff 2016: 133, Fig. 7). Underground drainage networks are rapidly forming along ice wedges in the Arctic today where climate warming is occurring (Fortier et al. 2007; Perkins 2007). Whether the gullies washed out during Late Pleistocene periglacial conditions, as suggested for channels in Belgium (Langohr & Sanders 1985), in Pleistocene yedoma (silt deposits) in Alaska ~40–25 ka (Douglas et al. 2011), or under warmer, more humid Holocene conditions as suggested for channels in Germany (Lang & Mauz 2006), is open to debate. Loose sandy fill within three deformed wedge structures has been dated to two periods, ~40–30 ka when gullying and fluvio-thermal erosion occurred, and ~17–13 ka when frost cracking occurred long after the initial gully fill (French et al. 2007: 52, Table 1).

Low-lying sediments would have considerably higher moisture content than adjacent uplands. Even in arid polar regions today, wetlands commonly occupy many low-lying surfaces (Tedrow et al. 1968; Woo & Young 2003; Woo 2012: 349). At riverine spungs, a near-surface water table would prevent wind scour from exceeding the coherent surface of such a table. This mechanism is similar to a Stokes surface (Stokes 1968; Fryberger et al. 1988), where the cohesion of wet unconsolidated sediments prevents wind scour near or at a water table (Fig. 4.20). The difference is that low-lying sediments in the periglacial Pinelands were at times frozen when frigid northwest winds eroded sandy ground to form spungs (French & Demitroff 2001: 346), although isolated pockets of unfrozen groundwater might have been present between dunes (Dinwiddie et al. 2014). The hard-packed flat bottoms have characteristics much like a fragipan, with little organic matter, high bulk density, cemented when dry, and low permeability (Witty & Knox 1989); yet spung bottoms lack a cementing of individual grains.

Where sediments are siltier, say, from loess deposition, wind erosion might form yardang (erosional remnants). At Schmidts Pond, a curious peg appears to be a yardang left in the center of a circular spung (Figs. 4.7, 4.21). West of Broad Pond flutes are wind-sculpted into silty sediments and densified paleochannel banks limit the blowout's width (Fig. 4.22).

Spungs: Summary

Spungs remain problematic. Although wind action was considered the dominant agent for spung formation (French & Demitroff 2001; Wolfe et al. 2023; also see Ambert & Clauzon 1992, FR basins; Carozza et al. 2016: 90, FR basins), other processes may have prepared basin sediments for deflation (e.g., nivation, piping, desiccation, thermokarst, vegetational changes). Spungs should be compared with other depressional wetlands, not only in the eastern United States (e.g., Delmarva Bays in DE, MD,

VA—Stolt & Rabenhorst 1987a; 1987b; Carolina Bays in NC, SC—Johnson 1942; Richardson 1981; Savage 1982; Moore et al. 2016; Citronelle Ponds in FL—Otvos 1973), but also with similar cold climate features in Europe (e.g., Mares in FR, DK—Cailleux 1961; Mardelles in FR, UK—Collin & Godard 1962; Viviers[21] in BE—Gillijns et al. 2005; Sölle in DE—Troll 1962; Crovuri in RO—Florea 1970; Gherghina et al. 2008). Spung extent, age, and groundwater association remain enigmatic in many ways.

---

21. The closed depressions on the Hautes Fagnes plateau in Belgium interpreted as pingo scars (Pissart 1965, 1974) and frost mound remnants (Pissart 2000, 2003) have recently been suggested to have an eolian component. Harris (2004) reports that the "Viviers are remarkably like deflation basins in arid areas," and interprets their ramparts as hairpin dunes (Harris 2004: 01253).

Fig. 4.1 Spung hydrology and intermittent fill.

(A) A typical medium-sized round spung on the Stockton University campus near the Arts & Sciences Building, by Oak Pond Road. Photo taken by Mark Mihalasky. Pomona, NJ. B) Early woodsmen aptly named South Jersey's enclosed basins as spungs or pocket bogs for their pouch-like form (Fig. 4.2).
(B) The above diagram shows a generalized form of a rounded spung and demonstrates the relationship between the groundwater table and the apparent water surface within it. During high periods of shallow groundwater, generally during winter and spring, pools of water fill these depressions through seeps and springs. When the shallow water table drops, generally during summer and fall, shallow basin bottoms lie above the local groundwater table and become dry land. Drawing by Pat Demitroff.

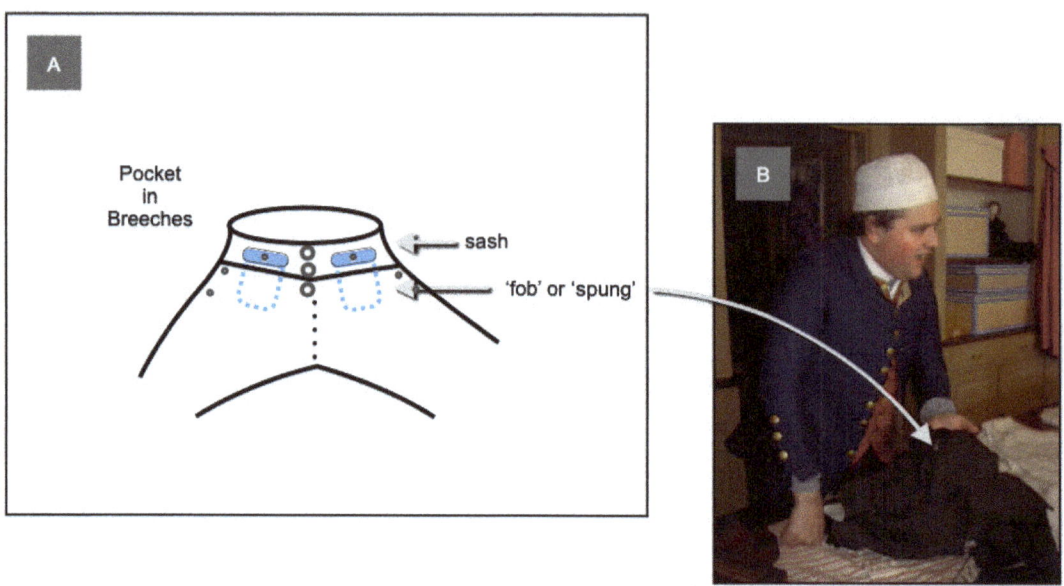

Fig. 4.2 Spungs, just pockets of water.

(A) An illustration of breeches, c. 1700–1750, showing fob pockets in the waistband (adapted from Cunnington 1972, Fig. 18). Pine Barrens spungs are water pockets. Reproduced with permission from Plays, Inc.
(B) Normally, a waistcoat (here red) hides the sash and its spungs from public view. Photo taken at the tailor's shop, Colonial Williamsburg, Williamsburg, VA.

Fig. 4.3 Horse Break Pond, a long-forgotten place.

New Rome, near Buena, NJ. Photo taken by Ron Bertonazzi and Mark Demitroff.

(A) From the air today Horse Break Pond appears irregular in shape but is a larger oval spung in earlier form. If stripped of its encroaching vegetation, its bowl-shaped bottom would be revealed, as might many other irregular spungs.
(B) At arrow, Wright (1867a) locates the pond on the Cohansey–Egg Harbor trail on Weymouth Furnace lands. This map is the only primary source found so far that names the feature.

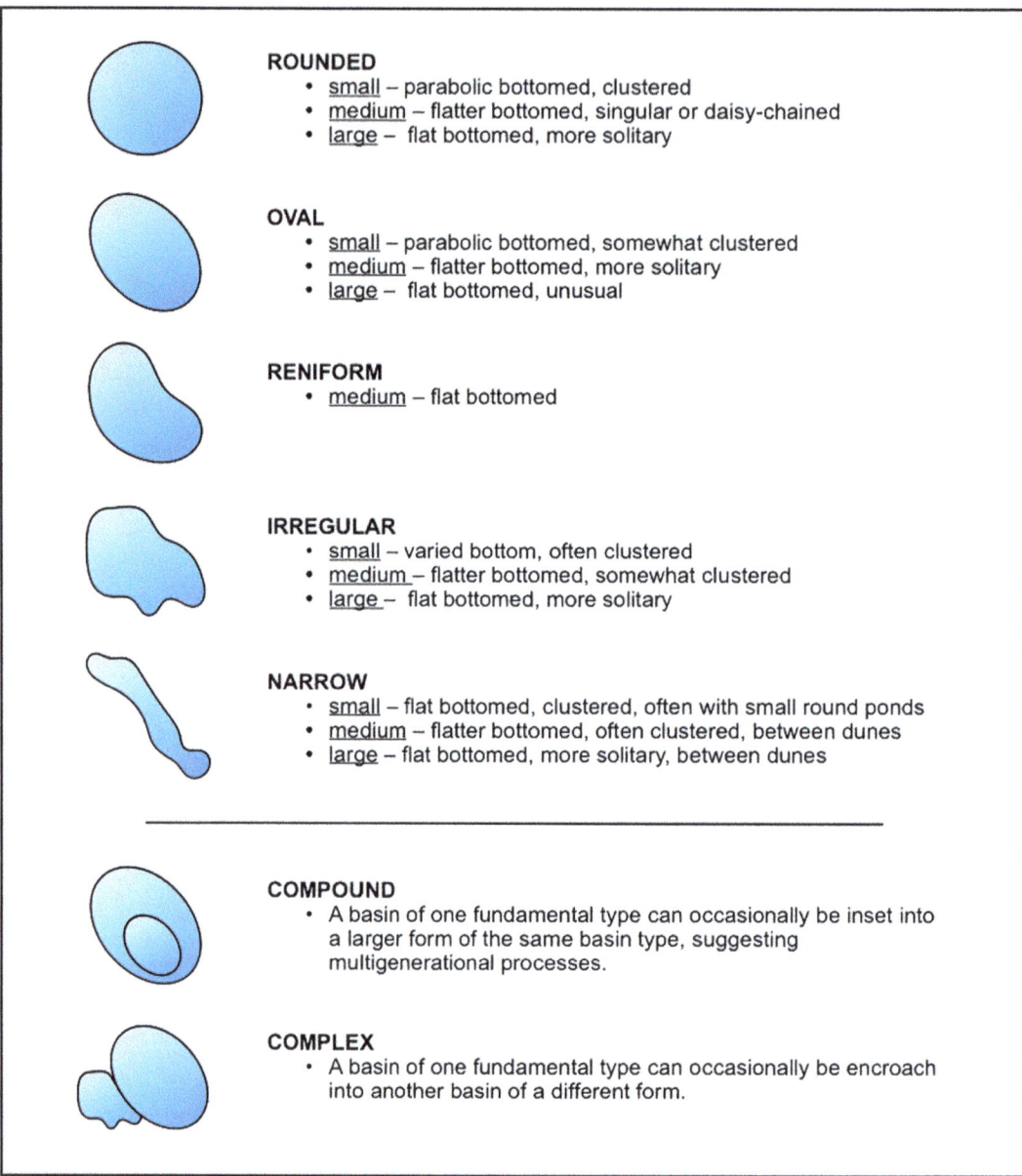

Fig. 4.4 List of spung forms observed in the Pine Barrens.

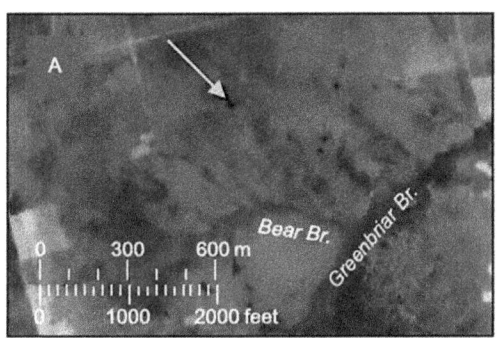
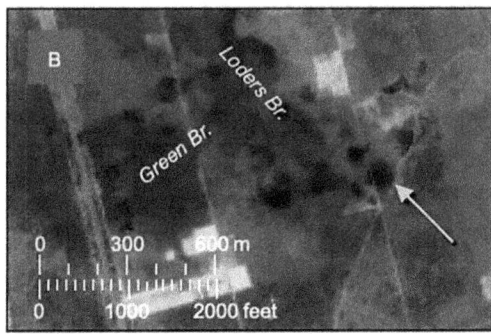

Fig. 4.5 Round Spung Group examples—small, medium, and large round spungs and their locations (at arrows).

Aerial imagery (NJOIT-OGIS c. 1931, accessed through 1930s Aerial Photography of New Jersey, Boyd Ostroff).

(A) Round Pond, Milmay, NJ.
(B) Lee Pond(s), above Doughtys Tavern, near Milmay, NJ.
(C) The Lake, Lake, between Piney Hollow and Malaga, NJ.

80 Soggy Ground

OVAL SMALL

Unnamed
- Thick 'n Hole, New Italy
- South River & Hance Bridge trails

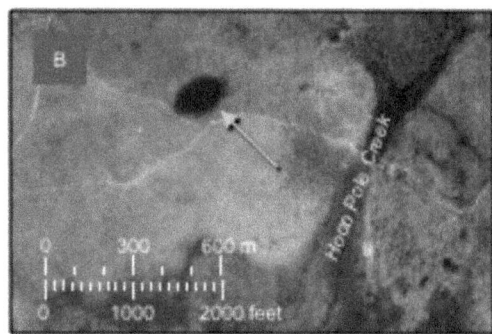

OVAL MEDIUM

Five Acre Pond
- Forgotten Ingersol

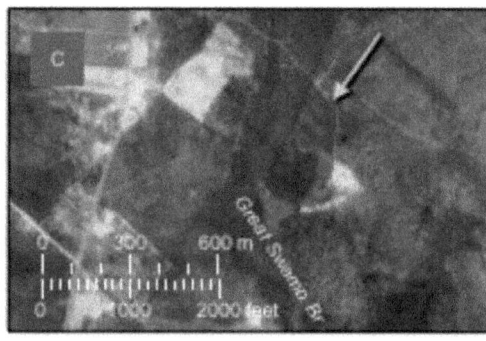

OVAL LARGE

The Oasis
- Tar Kiln Neck, DaCosta
- Maeroahkong trail

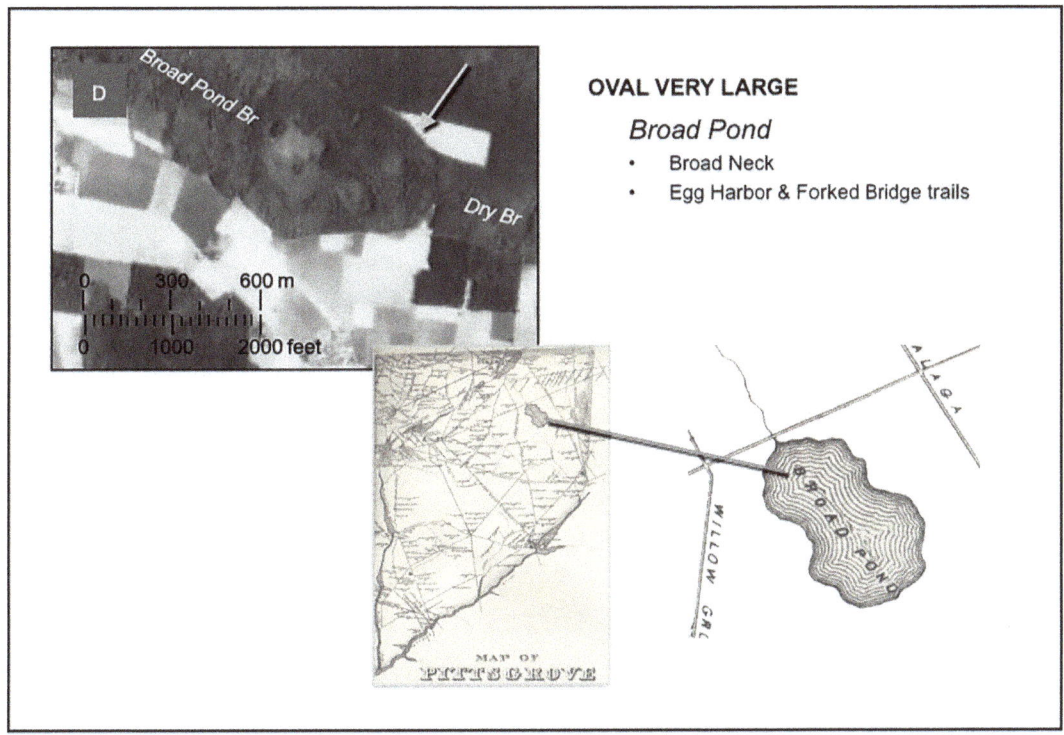

Fig. 4.6 Oval Spung Group examples—small, medium, large, and very large oval spungs and their locations (at arrows).

Aerial imagery (NJOIT-OGIS c. 1931, accessed through 1930s Aerial Photography of New Jersey, Boyd Ostroff).

(A) Unnamed, Vineland, NJ.
(B) Five Acre Pond, Estell Manor, NJ.
(C) The Oasis, near Hammonton, NJ. Another Oasis exits in New Italy, Vineland.
(D) Broad Neck, near Elmer, NJ. Map (Everts & Stewart 1876; 26).

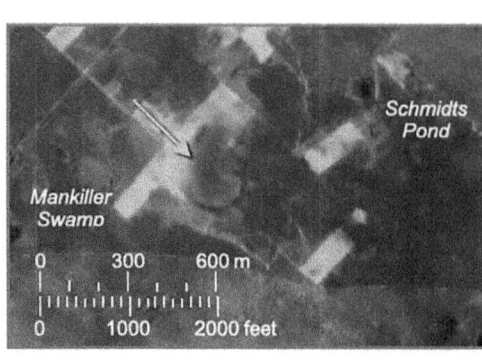

Fig. 4.7 Reniform Spung Group example—a medium reniform spung and its locations (at arrow).

Aerial imagery (NJOIT-OGIS c. 1931, accessed through 1930s Aerial Photography of New Jersey, Boyd Ostroff).

Kidney-shaped Winks Meadow appears to be the product of the coalescence of two round basins. Note two adjacent round basins to the immediate NW and NE of Winks Meadow, the latter being Schmidts Pond. South of Egg Harbor City, NJ.

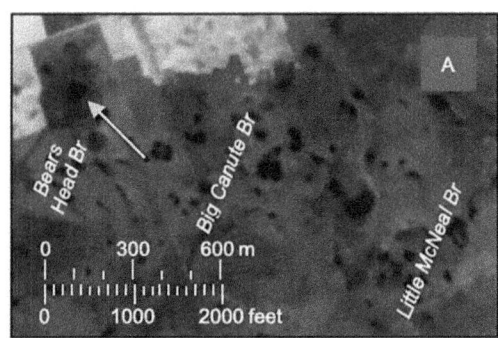

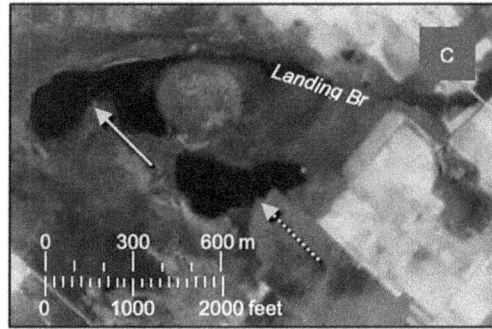

Fig. 4.8 Irregular Spung Group examples—small, medium, and large oval spungs and their locations (at arrows).

Aerial imagery (NJOIT-OGIS c. 1931, accessed through 1930s Aerial Photography of New Jersey, Boyd Ostroff).

(A) Bears Head Pond, Bennetts Mill, Vineland, NJ, one of at least four similarly named head ponds along Bears Head trail.
(B) Lummis Pond, Richland, NJ.
(C) Big & Little Goose Ponds, south of Egg Harbor City, NJ.

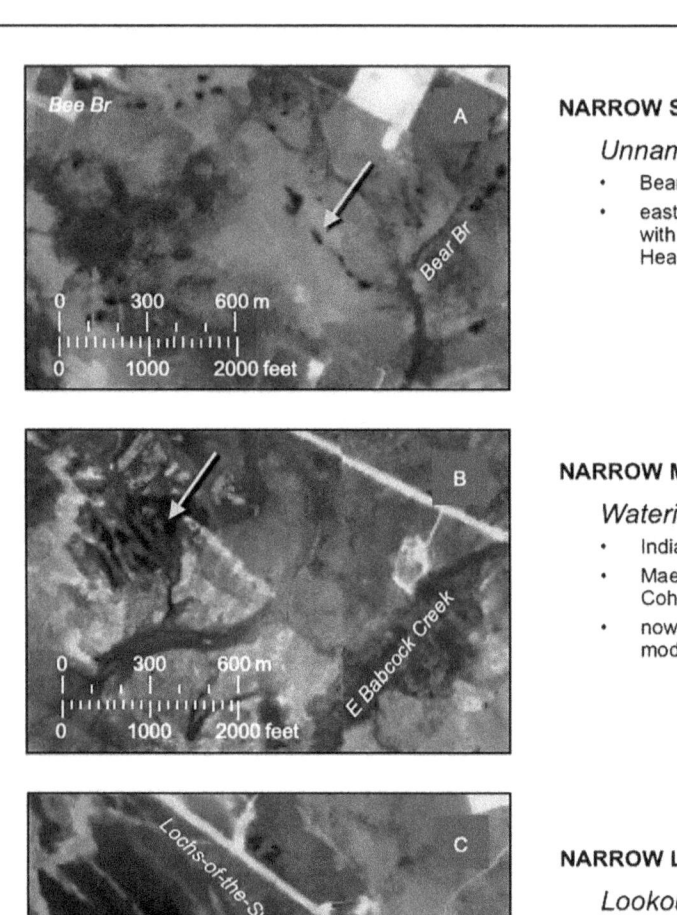

Fig. 4.9 Narrow Spung Group examples—small, medium, and large narrow spungs and their locations (at arrows).

Aerial imagery (NJOIT-OGIS c. 1931, accessed through 1930s Aerial Photography of New Jersey, Boyd Ostroff).

(A) unnamed pond, Bears Head hamlet, near Mays Landing, NJ
(B) Watering Place Ponds, Clover Leaf Lakes, north of Mays Landing, NJ
(C) Lookout & Crane Ponds, between Lauraldale & Mays Landing, NJ.

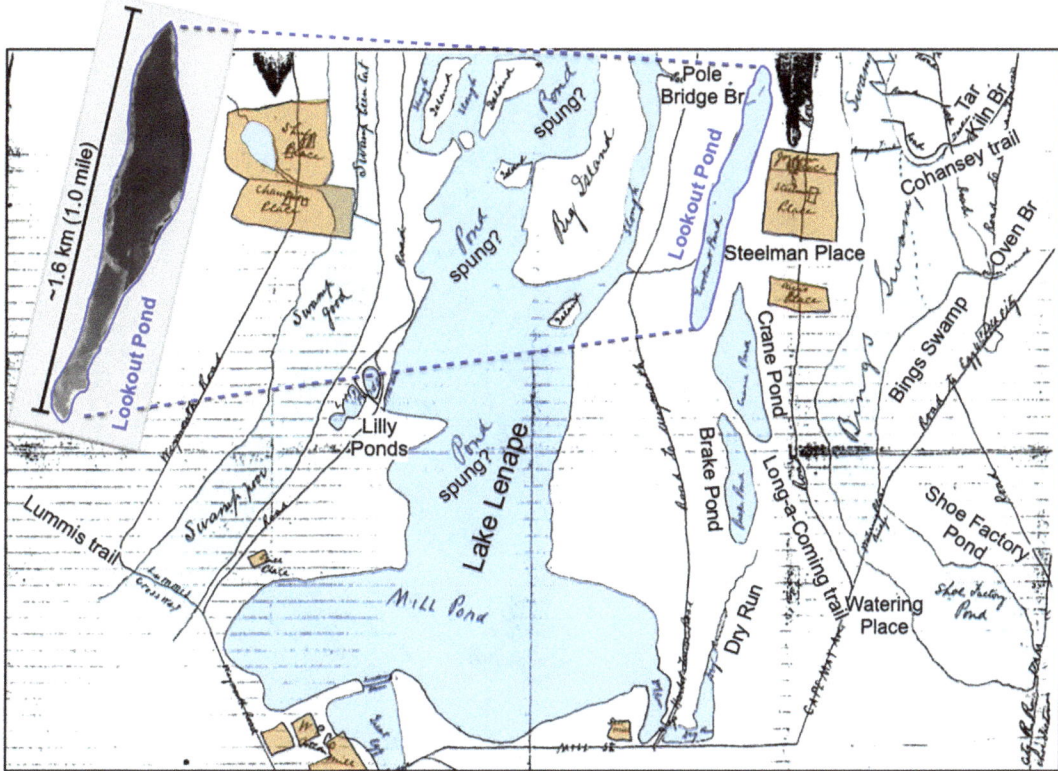

Fig. 4.10 Large narrow ponds at Great Neck.

The largest of the known narrow spungs were Lookout, Crane, and Brake Ponds (Wright? c.1867) at the Lochs-of-the-Swamp (Denny 1774) on Great Neck (Steelman 1741/1742). It was here at the junction of the Long-a-Coming and Cohansey trails that Mays Landing had its start prior to 1710 at the Steelman Plantation (personal communication, Jim Steelman, November 2003). The Lochs (Gaelic for a lake, esp. when narrow, OED Third Edition, December 2015) may have been named by Scottish tar kilners who settled in 1707 (Green c. 1920s) at nearby Nescochague, due north on Kings Road from the Steelman Plantation. Nesco was home to the geologist Peter Wolfe, who first proposed that the Pine Barrens was affected by Pleistocene permafrost (Wolfe 1952, 1953).

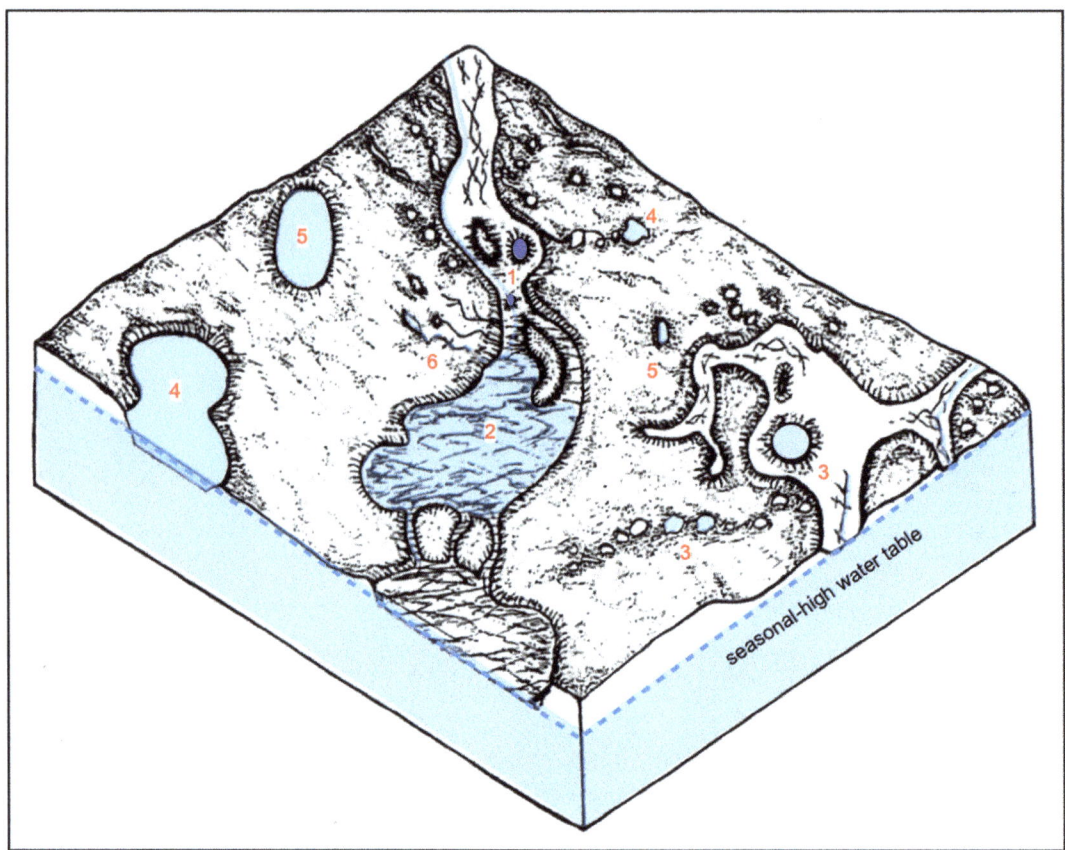

Fig. 4.11 Diagrammatic illustration of some pond types and settings.

The seasonal high water table level is idealized as planar to reflect the dynamics of a sandy surface deposit — 1) blue holes, which look spung-like but are not; 2) riverine spung; 3) round spungs; 4) irregular spungs; 5) oval ponds; and 6) furrows. Spungs are all ponds, but not all ponds are spungs.

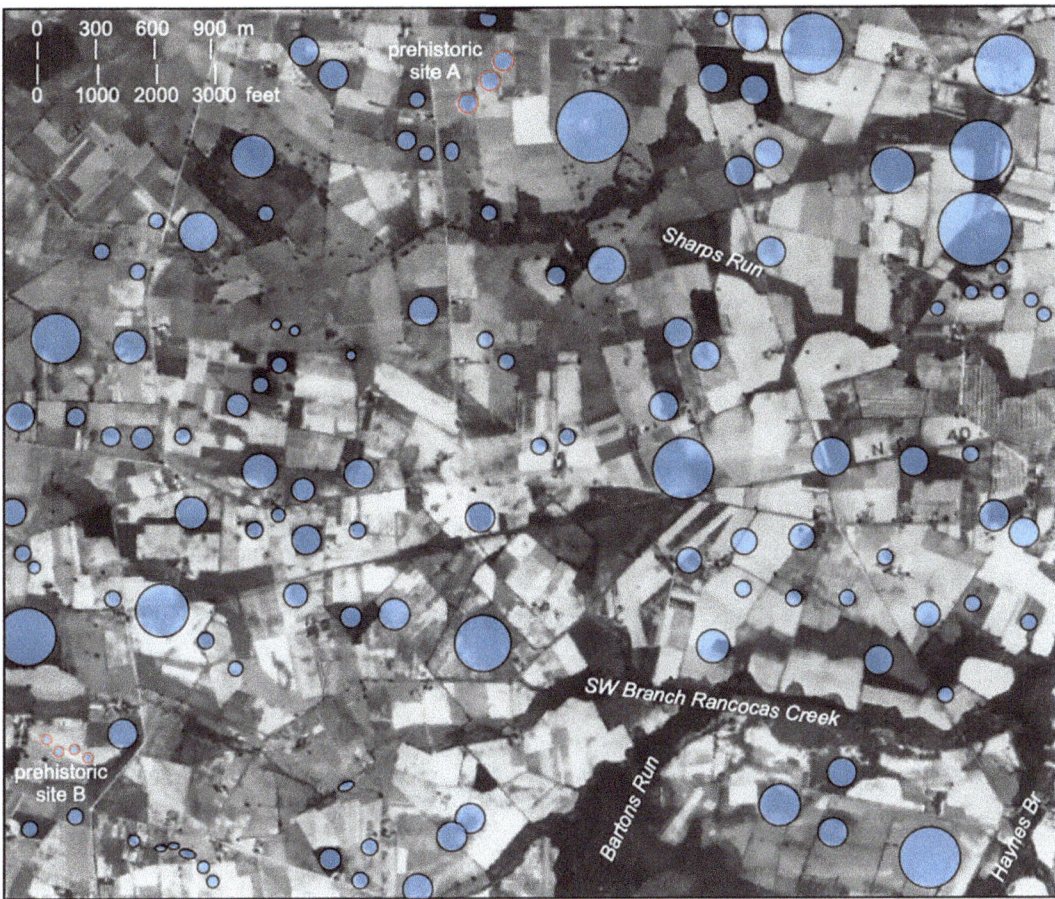

Fig. 4.12 Once abundant, basin sites are quickly disappearing.

Thousands of enclosed basins (blue circles) once dotted the Inner and Outer Coastal Plains of New Jersey. During the 1970s and 1980s, Cresson tested various spungs on the Inner Coastal Plain near Medford, NJ, for their archeological significance (e.g., prehistoric sites A & B). He found all contained evidence of significant precontact usage. Of the three known potentially intact Paleoindian sites found on the NJ Coastal Plain, all are associated with periglacial features. While two are in the Outer Coastal Plain, the last is in far easternmost Inner Coastal Plain in Lumberton Township and at present is at risk of destruction. The latter is one of the largest periglacial landforms surviving in this portion of Burlington County (personal communication, Jack Cresson, August 13, 2018). This figure is adapted from Cresson et al. 2006, Burlington County, NJ. Aerial imagery (NJOIT-OGIS c. 1931, accessed through 1930s Aerial Photography of New Jersey, Boyd Ostroff).

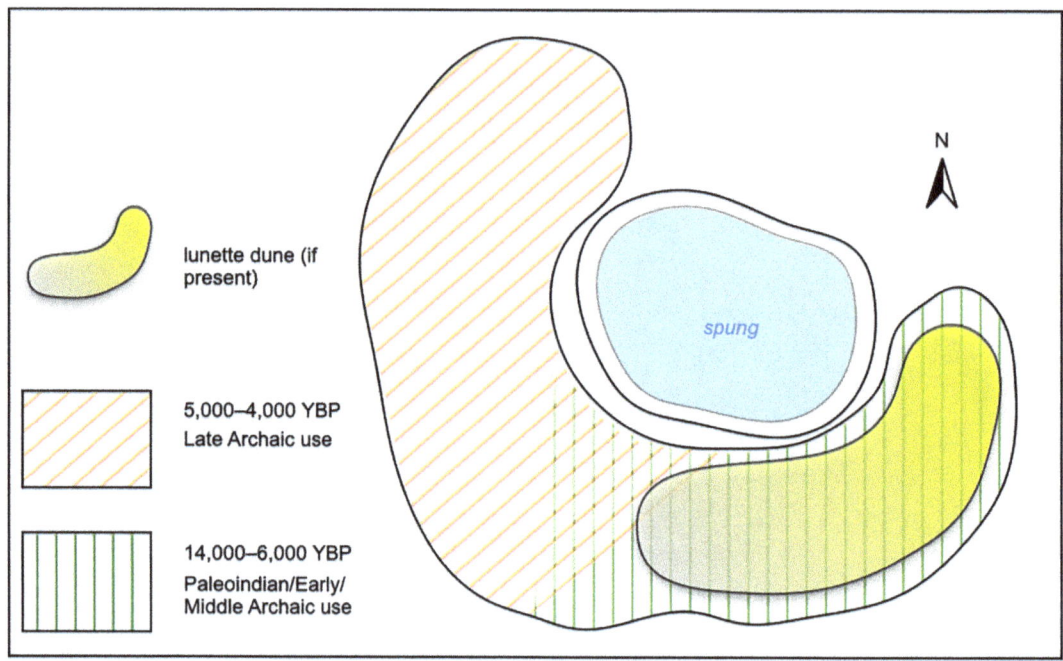

Fig. 4.13 Spung topography and precontact cultural use.

This diagram illustrates a generalized relationship between the lunettes of enclosed depressions in southern New Jersey and the precontact cultural artifacts found upon them. The evidence for prehistoric use of periglacial landforms (i.e., depressional wetlands) was first recognized in the 1960s and later published (Bonfiglio 1985; & Cresson 1978; 1982; Cresson 1983; Demitroff 2003). Early investigations showed marked patterns of Late Wisconsinan and Early Holocene aboriginal (Paleoindian, Early and Middle Archaic) adaptations with distinct settlement and resource exploitation. The earliest cultural use or activity areas were aligned with southern or southeastern uplands (i.e., rims) of basins, while later occupations clustered on south and southwestern sectors. Inferences and past megafauna discoveries in similar bog or lowland settings indicate the possibility that these basins were employed as "game traps." Based on historical accounts and documents, these unusual "watering holes" and wetland habitats were focal points for early historic travel (Cresson et al. 2006).

Fig. 4.14 Spung topography and trail configuration.

Ancient trails first linked water resources for animal, then aboriginal usage. Early settlers in the Pine Barrens later exploited these convenient byways. Aerial imagery (NJOIT-OGIS c. 1931, accessed through 1930s Aerial Photography of New Jersey, Boyd Ostroff).

(A) Swains Stopping-off Place was an early hostelry sited on the Old Cape Road, and home to the Methodist Circuit Rider, Richard Swain (Swain 1792). The old Cape Road, or Kings Road, followed Indigenous paths when laid out in the early eighteenth century (Elmer 1869: 73; McCahon 1994: 26). Survey excerpted from Hartman (1978: Map 8). The Swain Place was strategically sited at the junction of several paths that met at a cluster of spungs. Early travelers found spungs opportune locations for the refreshment of man and beast. Early taverns or "watering holes," and home sites were often established near such sites. Near Aetna Furnace, Head-of-the-River, NJ.

(B) Several trails met along the southeastern shore of Grassy Pond (Harshberger 1916: 143), a prehistoric pattern of passage mirrored by the course of the modern roads. The oldest, most serpentine iteration of the Tuckahoe trail (#1) hugged the east side of the Tuckahoe River and forded at Smiths Mill. A straighter route (#2) was laid with short bridging at Head-of-the-River. When a long drawbridge bridge was built at Tuckahoe (McCahon 1994: 120) in 1831–32 a straight route was laid out (#3). Grassy Pond remained an important waypoint throughout each iteration of road improvement. In 1910, strong springs emanated from the grassy bottom and charged this pond year-round with cool, clear water. Nearby residents could boat, fish and swim in this water body until the 1920s (personal communication, Edith Fanucci, September 1999). Today, Grassy Pond has overgrown with mixed hardwood forest with little trace of its former prominence. Between Dorothy and Milmay, NJ.

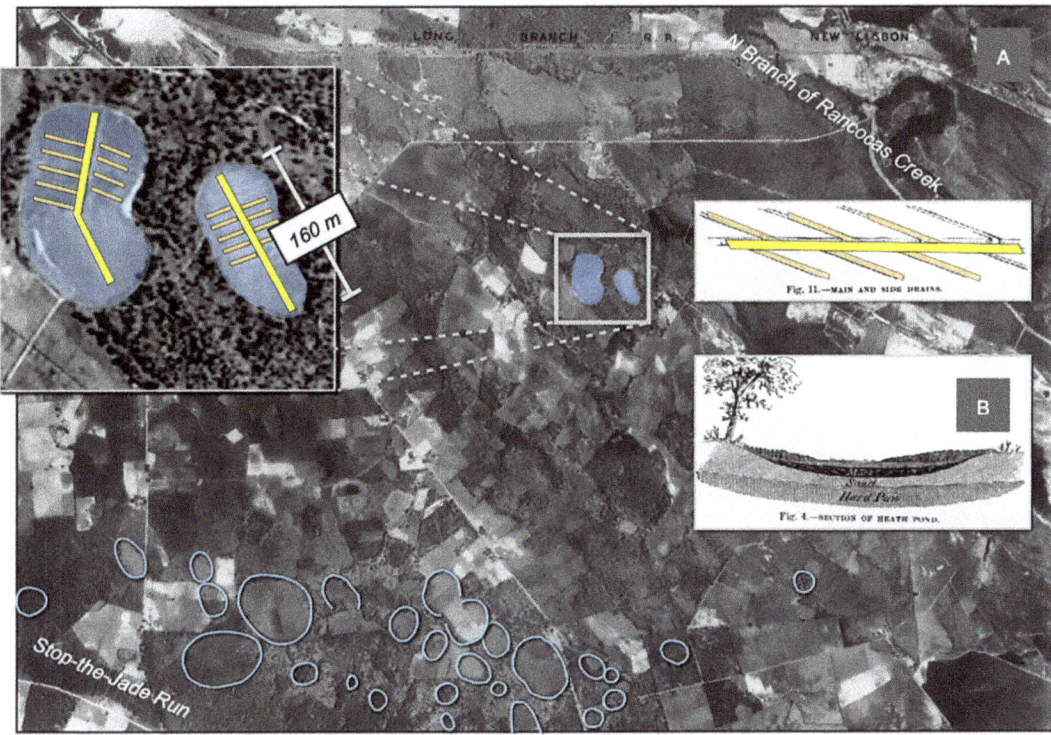

Fig. 4.15 Spungs and early cranberry exploitation.

(A) Joseph White began his cranberry industry career at Rake Pond tract, New Lisbon, NJ. There once existed a series of spungs—some outlined in blue. Aerial imagery (NJOIT-OGIS c. 1931, accessed through 1930s Aerial Photography of New Jersey, Boyd Ostroff).
(B) In his popular monograph on cranberry culture, White (1885: 29–30) touts the utility in the exploitation of Pine Barrens spungs that he called heath ponds. The expanded view shows early aerials water-control ditches in a configuration like those recommended in White's monograph. Excerpts (White 1885: 30, Fig. 4 & 38, Fig. 11).

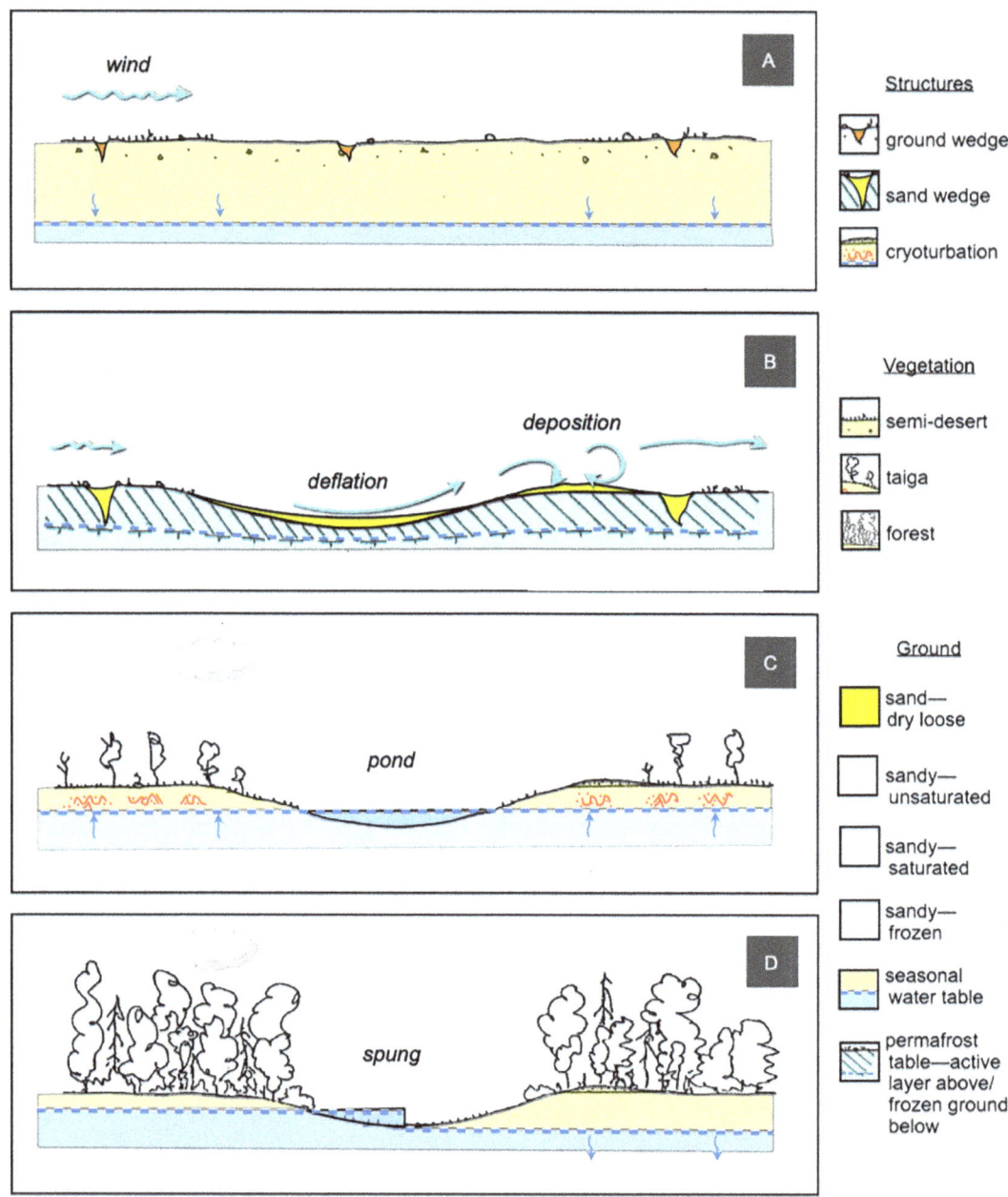

Fig. 4.16 A model of Pine Barrens spung evolution.

The seasonal high water table level is idealized as planar to reflect the dynamics of a sandy surface deposit. Modified from French and Demitroff (2001: 347, Fig. 5) to reflect the work of Wolfe et al. (2023).

(A) The original land surface of the Pine Barrens during the Late Pleistocene had been modified during earlier periglacial periods. Thirty thousand years ago the ice sheets are again in advance, the vegetation that appeared during the Mid-Wisconsinan some fifty to thirty thousand years ago becomes increasingly sparse.

(B) During the glacial maximum of the Late Wisconsinan about twenty-five thousand years ago, periglacial conditions (i.e., cold, dry, windy) prevailed. Katabatic winds swept across the sparse, desert-like landscape creating deflation hollows along lines of weakness (e.g., vegetation-poor patches, sediment-filled pots, furrows, thermal-contraction cracks).

(C) Permafrost began to thaw and the climate ameliorated so the shrub tundra and open woodland returned about nineteen thousand years ago. By fourteen thousand years ago the regional groundwater table rose and the depressions became intermittent ponds, some utilized by Paleoindians.

(D) A gradual lowering of the regional water table during the twentieth century has led to the progressive shrinkage and degradation of spungs and other shallow water-table dependent features in the Pine Barrens.

Fig. 4.17 Vignettes of the spung environment over time.

(A) Deflation could occur once the vegetational cover became semidesert (<30% sedges, mosses, and lichens), although during the coldest episodes little vegetation survived (Demitroff 2016). A helicopter provides scale, Banks Island, Northwest Territories. Photo courtesy of Hugh French.
(B) With the return of precipitation with warming, the local water-table rose to fill the ponds. Goldstream Valley, above Fairbanks, Alaska.
(C) Today, with a warming climate and over-withdrawal of groundwater spungs are quickly fading from the South Jersey landscape. At Egg Harbor Pond, only a slough (outlined in blue) dug into the spung bottom during the 1960s on-occasion reveals groundwater that once filled this renowned pond.

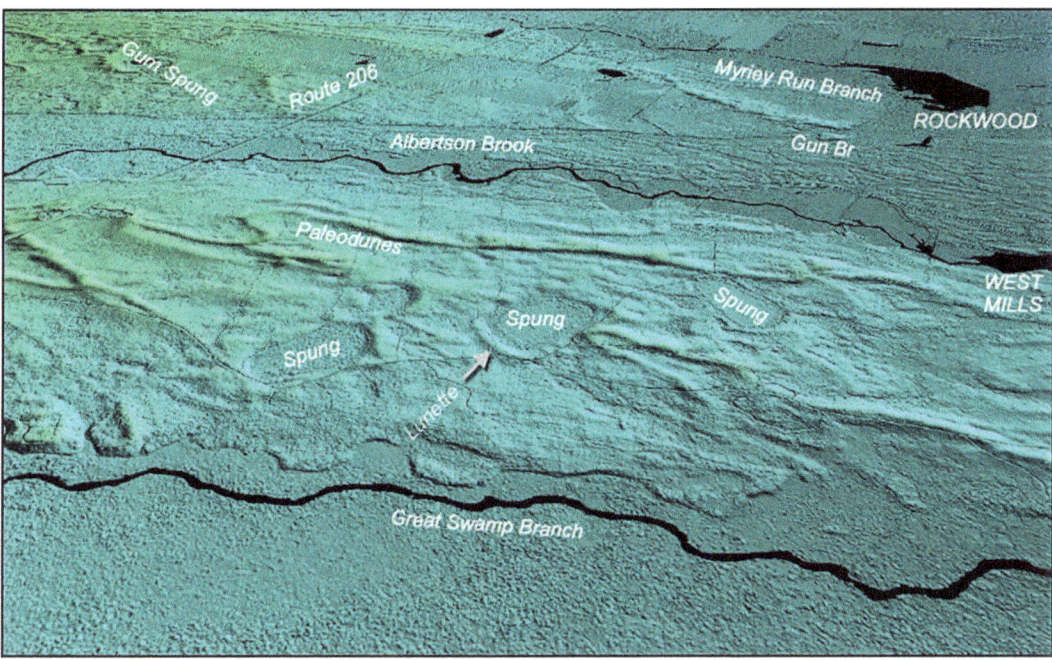

Fig. 4.18 West Mills dune field dynamics and spungs.

Here transverse dunes became hairpin parabolic dunes as vegetation returned about nineteen thousand years ago to finally end wind action to the exposed sands. Concurrent with dune building, strong katabatic winds generated deflation basins (spungs). An anchored dune called a lunette (at arrow) was built by the wind removal of basin floor sediments about 19,000 years ago (Wolfe et al. 2023). Some basins are older (Arsenault, in press) or much older relicts of earlier cold episodes. Lidar imagery (USGS & NJDEP, accessed as Lidar in the Pines 2022, Ostroff 2021).

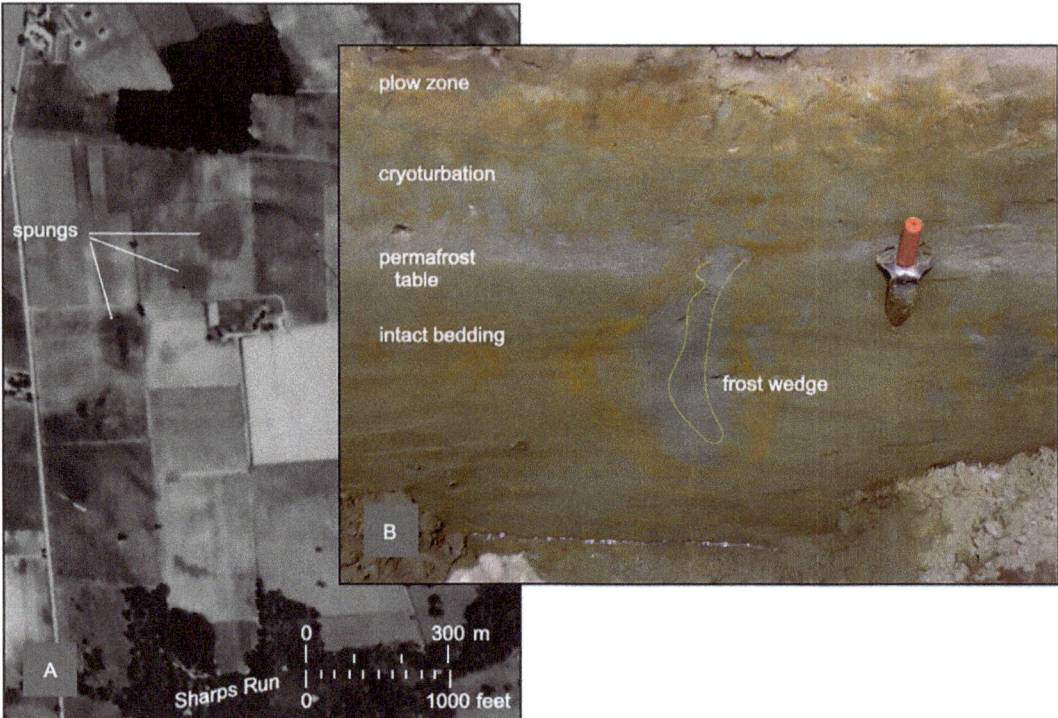

Fig. 4.19 An archeological study at a series of spungs.

(A) Three Inner Coastal Plain spungs near Medford, NJ, were investigated for their archeological significance (Tomaso et al. 2008). In genesis they were determined to be of deflational origin, Aerial imagery (NJOIT-OGIS c. 1931, accessed through 1930s Aerial Photography of New Jersey, Boyd Ostroff).

(B) Backhoe excavation across the basin features revealed a polygonal network of frost wedges of fine silty sand infill all along the trench walls. The upper parts of these wedges have been cryoturbated out of existence as the active layer deepened with climate amelioration. The wedge tails were presumably preserved (entombed) in frozen ground, which preserved not only the wedge-tail casts, but the original bedding of the spung infill. Woronko et al. (2022b: 6) report that frost cracks translocate water in a manner to agricultural tile drains, leaving behind a halo zone in sand-wedge casts not unlike the frost structures present here.

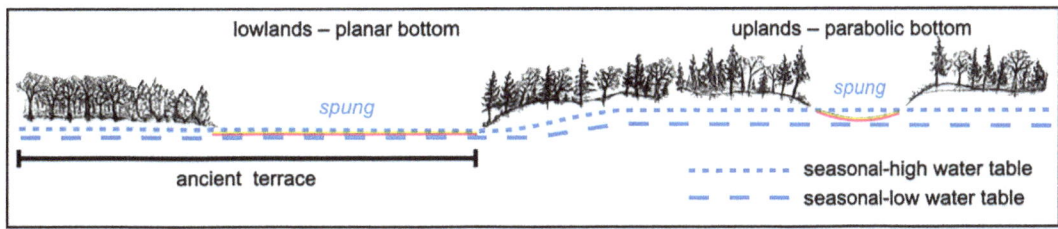

Fig. 4.20 Two distinct spung-bottom configurations.

Their difference in their floor architecture appears to be related to the depth of shallow groundwater during deflation. It remains unclear what state that water took—frozen or unfrozen at the time of sand entrainment. Elevations are exaggerated for illustrative purposes. Drawing by Pat Demitroff.

The first spung type (*on left*) relates to large ponds, typically with broad flat floors (e.g., Little and Big Goose Ponds, Fig. 4.8C). These are positioned in lower, wetter areas where groundwater was—and still is—likely close to the surface. Wet (frozen?) sediments positioned below the water table were unaffected by wind erosion (Stokes 1968, Stokes surface). Only loose, dry basin sediments could be entrained and only to the water table, not unlike playa like the Bonneville Salt Flats in Utah. This might help to explain the remarkably planar bottoms at the Little and Big Goose Ponds.

The second spung type (*on right*) has a parabolic floor and is the typical form for smaller ponds occupying higher and drier positions in the Pines. In elevated positions, sediments would have been low in moisture, allowing for parabolic deflation basins formation. The rounded lining of the first type is reminiscent of small deflation basins forming along the Dutch coast today (Jungerius 1984; & van der Meulen 1989). Arsenault (in press) reports ancient parabolic-bottomed spungs on the Outer Coastal Plain near Clayton, NJ, that now accommodate up to 3.5 m (12 feet) of pond sediments, their deposition beginning about 20,500 years ago.

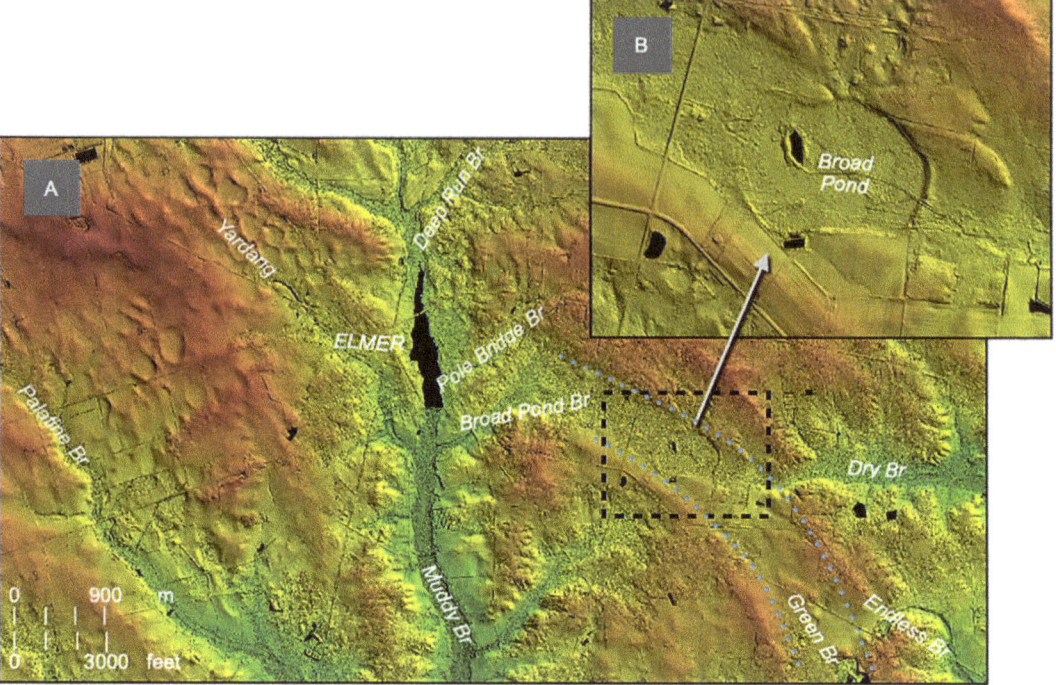

Fig. 4.21 A large oval spung formed within an ancient paleochannel.

Lidar imagery (USGS & NJDEP, accessed as Lidar in the Pines 2022, Ostroff 2021).

(A) Broad Pond, near Elmer, NJ, resides at the edge of the siltier Inner Coastal Plain, close to the sandier Outer Coastal Plain just to the east. One would not expect such a large oval-shaped spung to form in such an area. Instead, strong katabatic winds often carve these siltier sediments into mega-flutes that may be likened to yardang, which resemble flutes on cobble ventifacts (Demitroff 2016), only at a much larger scale.

(B) Recent access to high-resolution bare-earth lidar imagery has helped to reveal why Broad Pond exists. This large spung lies between the banks of a broad (cold climate?) ancient paleochannel that accommodates periglacial sandur deposits that are easily deflated. Its bounds are bracketed by the hardened banks of the ancient river channel and accounts for its breadth.

Fig. 4.22 Butte-type erosional remnant within circular spungs.

On occasion, a plug-like structure thought to be a yardang can be found near the center of a round spung.

(A) This phenomenon is best expressed in the Pines at Schmidts Pond, near Egg Harbor City, NJ (Fig. 4.7).
(B) In warm regions there are plant-pedestal yardang, as seen at White Sand National Park, near Alamogordo, NM.
(C) Similar cold climate structures are reported by Troll (1973: 6, Fig. 1) as rofbard in Iceland. Seppälä (2004: 144, Fig. 6.4) describes butte-type deflational remnants from Finnish Lapland. Day et al. (2016) describe peg-like yardang present within circular Mars craters, artifacts of wind action generated under periglacial conditions. Reproduced with permission from ©Schweizerbart'sche Verlagsbuchhandlung (Schweizerbart Science Publishers).

100   Soggy Ground

05
Cripples of the Pine Barrens

Etymology of Cripples

Cripples are obscure landforms that have barely been recognized or investigated in the Pinelands (Figs. 5.1–5.2). The meaning of "cripple" is confused and frequently associated with "spung" in modern usage, and in the literature (Weygandt 1940: 48–50; McPhee 1968: 61; Bisbee 1971: 287, 290; Neuendorf et al. 2005: 150, 621). Gifford[1] (1900: 247) applied the term to any brushy tangle and considered this to be a localism used in southern New Jersey and on the Delmarva Peninsula, although its usage occasionally extended into Pennsylvania and New York (Hall 2012: 852, E New York, E Pennsylvania, New Jersey). McCormick (1970) defined cripples as wooded swamps that "grow along smaller stream courses and intermittent drainages." This author believes that spungs and cripples are distinct features, each related to cold-climate geomorphology. Unlike the enclosed spungs, cripples have open drainageways. At certain times cripples can behave like rivers (Neuendorf et al. 2005: 150) but lack a significant incised channel.

---

1. John Clayton Gifford (1870–1949), North America's first PhD forester, grew up in the Pine Barrens village of Mays Landing and spent much time interacting with the local landscape (Troetschel 1950). He was founding editor of a trade journal *The Forester*, which gave a bully pulpit for conservation. One pet project of the editor and his forester wife (Edith Wright McCarthy Gifford) helped stop the destruction of the Palisades cliff wall threatened by quarrying operations for railroad ballast (Haffner & Mitchell 2019). This National Natural Landmark on the Hudson River is opposite New York City and is the western steep wall of a U-shaped fjord carved by a glacier.

The word cripple is derived from the Dutch "kreupelbos" (thicket, brake or underwood) and "kreupelhout" (underwood, undergrowth; Van Wely 1973), also cripple-bush from the Dutch kreuple-bush (Hall 2012: 852). According to a Dutch hydrologist from Vrije Universiteit, Netherlands (personal communication, J. J. de Vris, 2003), "kreupelhout" translates to "cripple wood," which are the bushes and underdeveloped low trees found under wet and marshy conditions in Holland today (Fig. 5.3). They usually consist of alder (*Alnus*), willow (*Salix*), and birch (*Betula*). Another possibility is that the feature is named for a type of "knott grasse"[2] that grew in Holland and the name was transferred to analogous herbaceous wetlands along the Delaware River (Harmar et al. 1860: 203). In Holland, coastal marshes and brook valley meadows were extensively grazed as meadows for many centuries (Bakker et al. 2002; Grootjans et al. 2002) as in Germany and other parts of Europe (Seer & Schrautzer 2014).

Dutch and Swedish colonists settled southern New Jersey and adjacent Delaware early in the seventeenth century. They used the term "Creupel" to describe short wetland corridors, a term which later became Anglicized to "Cripple." A grant given in 1666 to John Ogle included 300 acres (121.5 ha) known as Muscle Cripple, "a piece of valley or meadow ground" (Delaware 1903: 157). Another early description (1676) of a cripple relates to a right-of-way for livestock passage in colonial Delaware. A landowner employed Martin Garretson to make a passageway over a "Valley & Creuple; but could not make ye sd way Sufficient for Cattle to goe over; by Reason of the Rottenness of ye ground, being a Quaking mire wch hath noe foundation for a way . . ." (Fernow 1877). This establishes that a cripple occupies a valley and is wet. In a 1680 court case in Pennsylvania, a cripple was equated with swamp (Harmar et al. 1860: 178). A map survey at the Millville Historical Society shows "lands, swamps and cripples" along the Maurice River (Miller 1749), and again intimates that cripples are stream-like. "Cripple" is frequently employed in land descriptions during the eighteenth and nineteenth centuries in association with elongated swampland (Delaware 1845: 34, 183, "meadow, marsh and cripple," 183 "meadow, marsh or cripple") (Fig. 5.4). The popular nursery rhyme, "The Farmer in the Dell" takes place in such a valley.

---

2. Knotgrass is Prostrate Knotweed (*Polygonum aviculare* [L.]). Its appearance in seventeenth-century New England is associated with the introduction of cattle (Del Tredici 2010: 254) and indicates overgrazing of meadows (Boonman & Mikhalev 2005: 389). There is some confusion as to the botanical nomenclature of *P. aviculare* as applied by Linnaeus (McNeill 1981).

Apellatives of Cripples

Rarely do old deeds, surveys and road returns give these multitudinous small drainageways a specific name. Usually, small ephemeral channels are simply called a "cripple," especially if their bottoms tend to be planar. Several named cripples include: Buck Cripple (Miller 1781; Clark & Hitchnor 1802) (Figs. 5.5, 5.10); Cedar Creek Cripple (Zinkin 1976: 52); Big Cripple (Farr 2002: 14); Birch Cripple (Clement 1845: 66); and Long Cripple (Bisbee 1971: 137). Cripples with concave-up bottoms are often referred to as hollows, hollers, or wallows. Hell Hollow was a miserable place where people were too poor to use candles (Pierce 1957: 56–57). Ten Mile Hollow is only two miles (3.2 km) in length;[3] 10 miles (16 km) is the distance to the post office (Zinkin 1976: 165). Others are Frog Holler (Zinkin 1976: 76), Grassy Hollow (Zinkin 1976: 80), Hog Wallow (McPhee 1968), Stob Hollow (Bisbee 1971), and Horse Wallow (personal communication, Robert Francois, 2003) (Figs. 5.12).

Cripple Morphology

In the low relief of the Pine Barrens, cripples are subtle open drainageways that behave like rivers, but lack significant modern incised channels. They are neither streams, nor swamps, nor intermittent pools (Fig. 5.6). Yet they appear to support important and sensitive habitat for a wide diversity of grasses, sedges, and orchids (personal communication, Mike Hogan, September 23, 2006). Their recognition fills an important gap in the characterization and protection of Pinelands wetlands (personal communication, Sandra Hartzog-Bierbrauer, October 2004). Little is known of their numbers, their extent outside the Pinelands, their age, or their importance concerning threatened or endangered plants and animals.

Three distinct forms are present in the Pinelands (Fig. 5.7). Two types (Fig. 5.8) are short, rather broad, spur at right angles to a larger valley, and lack a well-defined, incised river channel. A longer form of cripple appears to be more stream-like (i.e., well defined U-shaped channel,

---

3. Place names like Five Mile Branch and Seven Mile Tree on the Weymouth–Forked Bridge Road relate to their mile-mark along that road. In this example, the former is but 3 km (2 miles) in length. However, eastward that branch is 8.85 km (5.5 miles) from the road's intersection with the early Woodbury Road. Going westward, Five Mile Branch is 3.3 km (2.0 miles) away from Seven Mile Tree on the same road.

sinuous path), yet surface flow is only intermittent and is sheet-like (Figs. 5.9–5.10). Long-form cripples often have subsurface water percolating through sand below their streambeds (Figs. 5.11–5.13). Planar-bottomed cripples tend to occupy areas of low relief, and are themselves rather shallow, usually about a meter (a yard) in depth. Cripples with concave bottoms are more common where the surrounding terrain is relatively elevated and have channels that can be several meters in depth. Some cripples are planar-bottomed towards their outlet but are more concave bottomed further up.

Most cripples discharge onto broad Late Pleistocene stream paleochannels in the Pinelands (Fig. 5.8). These short passageways are likely of age and origin similar to the broad paleochannels since cripples outlet at the same level as Late Pleistocene stream terraces. Jahn (1956: 450) uses similar reasoning to deduce relative dates for "denudation valleys" (periglacial dells) of the Lublin Plateau, Poland. Woronko et al. (2022a) report that U-shaped periglacial dells in northeast Poland developed their distinctive shape during the extreme climate fluctuations of the Mid-Wisconsinan some 60,000 to 30,000 years ago.

Valley asymmetry, attributed to insolation (exposure to the warming sunrays) and periglacial processes, has been noted along larger riverbanks in South Jersey (French et al. 2007).[4] In Europe, short valleys often exhibit slope asymmetry (Maarleveld 1949, 1951; Johnsson 1961; Czudek 1986). Maarleveld (1951) and Dzieduszyńska et al. (2023) suggested that cold-climate processes may account—at least in part—for dell asymmetry in the Northern Hemisphere, controlled by snowmelt-driven solifluction in Western Europe and by insolation in North America. Cripple delineation and Geographical Information System (GIS) analysis could be used to determine the extent of channel asymmetry present in the Pinelands.

---

4. During frigid episodes the ground remained frozen late into the summer (deep seasonal frost) or remained frozen all year round (permafrost). Although air temperatures were quite cold, the sun's rays were still warming. Radiant energy was strong enough to thaw the ground in direct exposure to the sun, especially when snow cover was thin. North- and east-facing slopes are often steeper here because they received the least amount of thawing sun. South- and west-facing slopes received the most amount of radiant energy, so they thawed and wasted away much more quickly than North- and east-facing slopes. This effect is not very well pronounced in the Arctic since there the sun's angle is very low and not very intense. That is not the case in the Pine Barrens. It is believed that in frozen mid-latitude South Jersey the resultant slope asymmetry is well expressed because of insolation—e.g. bluffs along the Great Egg Harbor River west bank from Weymouth to Mays Landing.

Early Cripple Use and Topography

Cripples are wetlands and support important plant and animal habitat. Not much is known of their cultural utilization. Wetter bottoms were ideal locations for gathering of Sphagnum moss and peat. The collection, drying, and baling of moss and peat for horticultural use was an important industry in the Pines (Waksman et al. 1943; Moonsammy et al. 1987: 140–143).

The edges of small, cripple-like wetlands were preferred locations for tar kilns, like the "old tar kiln" on Tar Kiln Branch (Clement 1867: 25—Milmay; Denny 1774, "old" by 1774!), one of various similarly configured Tar Kiln Branches (Hartman 1978: Map 5—Pancoast Mills, & Map 6—Vineland; Wright? c. 1867—Carmantown). Naval stores (e.g., tar, pitch, turpentine) were rendered from pitch pine in these kilns. Resin-rich knots and stumps were collected and piled into large saucer-like enclosures. The gathered "fatwood" was then covered with turf and burned in a manner reminiscent of charcoal making (Fry 1918; Blinn 1918; Weiss & Weiss 1965; Wrench 2014) to drive out liquid constituents. It took great skill to slowly fire the pine sections from top to bottom, yet not consume the flammable resin collecting in gutters below. The drop in relief aside a cripple wetland provided a convenient gradient for sap collection (Blinn 1918: 29; Weiss & Weiss 1965: 27; Mounier & Wilson 2006). Another benefit of the location would have been the abundance of dense, peaty, and moist root mats for turfing the outside of the kiln. Sods made of this material would have provided excellent air-infiltration control needed during combustion.

Origin of Cripples

Wolfe (1941, 1943) acknowledged the existence two small valley types in central New Jersey. He called very shallow strike valleys "vales" after Woodbridge and Morgan (1937: 203), and broad, round-headed depressions "dales" after Thornthwaite[5] (Wolfe 1943: 204). Wolfe's area of study

---

5. Thornthwaite, through critique, helped mold Wolfe's Ph.D. dissertation (Wolfe 1941: 11). Mather and Sanderson (1996: 12) state that Geographer Carl Sauer had a great influence over Thornthwaite's work. They state, "At Berkeley, Sauer taught an introductory course in physical geography ... and it was important ... that climate, rather than W. M. Davis's concepts of geomorphology, formed the nucleus of this course." Thornthwaite (1899–1963) was Sauer's (1889–1975) student, and in turn enlightened Wolfe during several field trips on the importance of climate to geomorphology.

was just 65 km (40 miles) outside the Pinelands on the non-glaciated landscape of the Hunterdon Plateau, proximal to Laurentide moraines. He attributed the genesis of two valley types (dales and vales) to "unusual conditions of erosion [that] may well have been determined by the tundra climate of the glacial period, sparse vegetation, and severe wind and fluvial erosion" (Wolfe 1943: 209). He also recognized numerous closely spaced gullies whose depth of erosion was limited by a dense "hardpan" (i.e., fragipan) in central New Jersey.

Demitroff (2003) suggested that cripples are analogous to the widely described features: *dellen* (named by Schmitthenner 1926; Büdel 1982); *dalen* (Maarleveld 1949); *daler* (Johnsson 1961); *dry valleys* (Bull 1939; French 1972; Dzieduszyńska et al. 2023); *vallees seches* (Klatkowa 1965); *vales* (Hamelin & Cook 1967); and *dells* (Smolová 2002; French 2018: 364); Dzieduszyńska et al. 2023) of Europe. Dry valleys attributed to periglacial processes are reported to be common in Long Island (Das 2007), although caution should be used in their interpretation since they occur on old outwash plains (cf. Woronko et al. 2018).

In Europe, dells are considered relicts of former periglacial conditions (Dylik 1952; Washburn 1979: 253; Büdel 1982: 305; French 2018: 364; Woronko et al. 2022a). Permafrost formed repeatedly in South Jersey during the Late Pleistocene (French & Demitroff 2003; French et al. 2003, 2005, 2007). Surface wash from upland catchment areas would create broad, flat drainageways over the impermeable frozen ground. Intermittent streams would have lacked the mechanical energy needed to erode frozen sediments (Kuchukov & Molinovsky 1983).

During periods when the ground was frozen, erosion would occur along lines of preferential thaw (i.e., fissures from frost and desiccation cracking), especially during the Mid-Wisconsinan (Woronko et al. 2022b). Deformed sand and ground wedges in the Pine Barrens are widespread. Their modification is attributable to fluvio-thermal erosion, slumping, piping, and sediment re-deposition during the Mid-Wisconsinan (French et al. 2005, 2007). In the loose, sandy/gravelly coastal plain sediments of Spitsbergen (Repelewska-Pekalowa 1996: 162–166), piping (tunnel erosion) and thermoerosion (thaw of ice by flowing water) create gullies that are widened by erosion after brief summer rains and snowmelt (i.e., nival processes) (Fortier et al. 2007; Levy et al. 2008). Spring discharge under periglacial conditions also affects the shape, width, and orientation of dells (Paul 2014). Short (to several 100 m [330 feet]), flat-bottomed channels often form, especially where gravels armor valley bottoms. A similar mechanism is suggested for the formation of Pinelands "cripples."

When permafrost thaws, the surrounding "thirsty" sands and gravels of the Pine Barrens intercept much of the surface flow. What reaches a cripple's catchment today, without a permafrost-enhanced perched water table, is insufficient to incise a stream into the abandoned channel bottom. Thus, the stream network becomes almost inactive (i.e., a dry valley) without a frozen ground perched barrier to water infiltration (a cryogenic aquiclude or aquitard) (Vandenberghe & Woo 2002: 491).

It is unlikely that cripples could form under current climatic conditions (Woronko et al. 2022a). In the High Arctic today, snowmelt occurs over a two-to-three-week period, releasing large quantities of water stored during winter into streams and wetlands (Woo & Young 2006). Although deep thaw is often associated with wet tundra soils such as found in large river valleys (Nelson et al. 1997: 373) and other perennially wet locations in the Arctic (Nelson et al. 1998), the depth of thaw above permafrost in ephemeral wetland environments is usually shallow (Woo & Young 2006; Woo 2012: 366).

South Jersey's Pleistocene soils were generally dry, except for low-lying areas that briefly collected rainwater and snowmelt in late spring or early summer. Icy permafrost could develop in these catchments, despite the well-drained nature of their enclosing sands and gravels. Coarse-grained sediments are not necessarily ice-poor as commonly supposed (Ballantyne 2018: 60; French 2018: 122–123). In temporarily wet locations, the presence of ground ice would consume much of the latent heat needed for deeper thaw to occur (Woo & Xia 1996), especially if a dry peat layer is present (Woo & Young 2006: 434). Similarly, the active layer would have been fairly shallow in South Jersey's ancient ephemeral wetland systems, needing more latent heat for thawing than in higher and drier spots (Woo & Young 2003: 297). French et al. (2005: 176; cf. Vandenberghe et al. 2016: 142) suggested lenses of ground ice were present in the Pine Barrens sandy-gravely sediments during extended cold periods because of the presence of relict thermokarst features like modified wedges, sediment-filled pots, and cryoturbation.

Preliminary work on sediments at the ephemeral Lee and Debbi Ponds (i.e., spungs) support the suggestion that a thin active layer was present in Pinelands' wetlands (Demitroff et al. 2007; French et al. 2009b). Soil minerals affected by freeze-thaw cryogenic disintegration were found within 0.3–0.5 m of the Pine Barrens land surface. This relationship between the physical weathering of soils and intense periglacial freeze-thaw processes has been extensively investigated in Russia (Konishchev 1982; 1997; & Rogov 1993; 2017; et al. 2005).

Water running over permafrost and groundwater sapping—while significant geomorphic agents—are not the only factors at work in shaping the cripples (Woronko et al. 2018; 2022a). Mass-wastage or corrasion (Dylik 1952: 10; Jahn 1975: 172) and wind action by deflation (Dylik 1952: 12) in a manner similar to the spung wind-action hypothesis (French & Demitroff 2001) both played a part in cripple genesis. Corrasion (mechanical erosion of solid materials moved along by wind, waves, running water, glaciers, or gravity (Neuendorf et al. 2005: 145) may have been dominant in the formation of convex-bottomed cripples, and groundwater sapping for straight pattern, low order corridors (Das 2007). Hall (1997) suggested that dells might be of animal-made (zoogeomorphic) origin, when ice-wedge troughs degrade as muskox frequent them for stream access. Washburn (1979: 253) cited dangers associated with the invocation of permafrost for dell formation. Three alternatives must be eliminated first: 1) solution of limestone; 2) greater precipitation in the past than today; and 3) stream beheading. Only the latter point of headwater capture by another stream is relevant to the Pinelands cripples.

Cripples: Summary

Short valleys, reminiscent of periglacial dells, abound in the Pinelands. Too numerous to name, save the occasional generic "cripple" designation, they are not branches, runs, or brooks. Early woodsmen who named these features were not lexicographers or geomorphologists, so their application of cripple is expectedly flexible. Spung-cripple chimeras do exist and defy clear-cut designation. So, what are they? In place of the English-based vales, dales, and dells, "cripple" is employed in a distinctive Pinelands parlance—a remnant of the region's seventeenth-century Dutch roots.

"Birch Cripple," occupied by the native gray birch (*Betula populifolia* Marshall), was noted in Burlington County by Clement (1845: 66). "Maple Cripple" has been recently recognized as a Pine Barrens wetland ecological community by botanists J. Arsenault and T. Gordon (Cook College Office of Continuing Professional Education Programs 2006: 7). Moist bottomland in the Pine Barrens supporting oak are known as "white oak bottoms"[6] (Harshberger 1916: 10, 104, 151). These are old stream channels—

---

6. Clatford Bottom (Small et al. 1970) is a periglacial dell containing a periglacial rock stream in England. The feature is full of folklore, including stories of a resident devil.

often cripples—infilled with Pleistocene sand-sheet sand. The presence of sandy deposits raise the ground just high and dry enough above the old channel floor to allow the establishment of oaks. The ground below the old stream channel floor is often deeply weathered too at depth, perching its shallow groundwater. In the odd wet years oak bottoms flood becoming intermittent drainage channels. With renewed awareness of the features, there is a need to record the flora and fauna found in this habitat. Although peat[7] collection and naval store production have been associated with cripples, the full extent of their cultural dynamics remains virtually unknown.

Careful lithostratigraphic and biostratigraphic work is needed to confirm that such valley systems, as in Europe, have periglacial origins. Cryolithogenic analysis may be particularly useful in identifying cold-climate stratigraphic components (Demitroff et al. 2007; French et al. 2009b; Wolfe et al. 2023). Wilkinson (2003) calls into question paleoenvironmental interpretations of some dry valleys in southern England, where anthropogenic processes had greater influence on dell formation than previously thought. Optical dating of sediments (Woronko et al. (2022a) could be helpful to establish a chronology of eolian episodes, especially pertinent to understanding long-type sand-choked cripples.

Like Pine Barrens silcrete, the rock-stream's lithics are also pitted with opferkessel that fill with water. However, at Clatford the rock dimples' water-fill will mysteriously drain at night, having been sucked dry by that demon (Wills 2014: 47–48).

7. Mucky organic Pinelands sediments are called "peat" out of practicality, but it is not really peat ("true peat" of Waksman 1942: 38, autochthonous). It may be better termed deep organic muck. Often there is *Sphagnum* in the matrix of Pine Barrens muck, but so are grasses, sedges, branches, pine needles, silt, and the kitchen sink. True peat is mostly *Sphagnum*, and northern peat is different from the Pinelands peat.

Fig. 5.1 A typical concave-bottom cripple in the Pine Barrens.

Once intermittent. this stream corridor is now dry. It was identified as a nameless cripple on an old survey (Wright 1867b; see Figs. 5.4, 5.11). Emmelville, NJ.

Fig. 5.2 A typical planar-bottom cripple in the Pine Barrens.

This nameless watercourse empties onto the Great Swamp Branch terrace (i.e., Pleistocene paleochannel), and is the right-most cripple in Fig. 5.9. Hammonton, NJ.

Fig. 5.3 A typical kreupelhout, Grootegast Village, Netherlands.

This municipality is named for the "gast" or "gaast," an elevated sandy ridge in an otherwise swampy area (van Wely 1973). The region was affected by periglacial conditions during the Weichselian (=Wisconsinan). River systems in this area accommodate relicts of fluvio-periglacial processes (van Amerom 1990: e.g., ancient icings, braided channels, dunes). Photo by Joop Klompien.

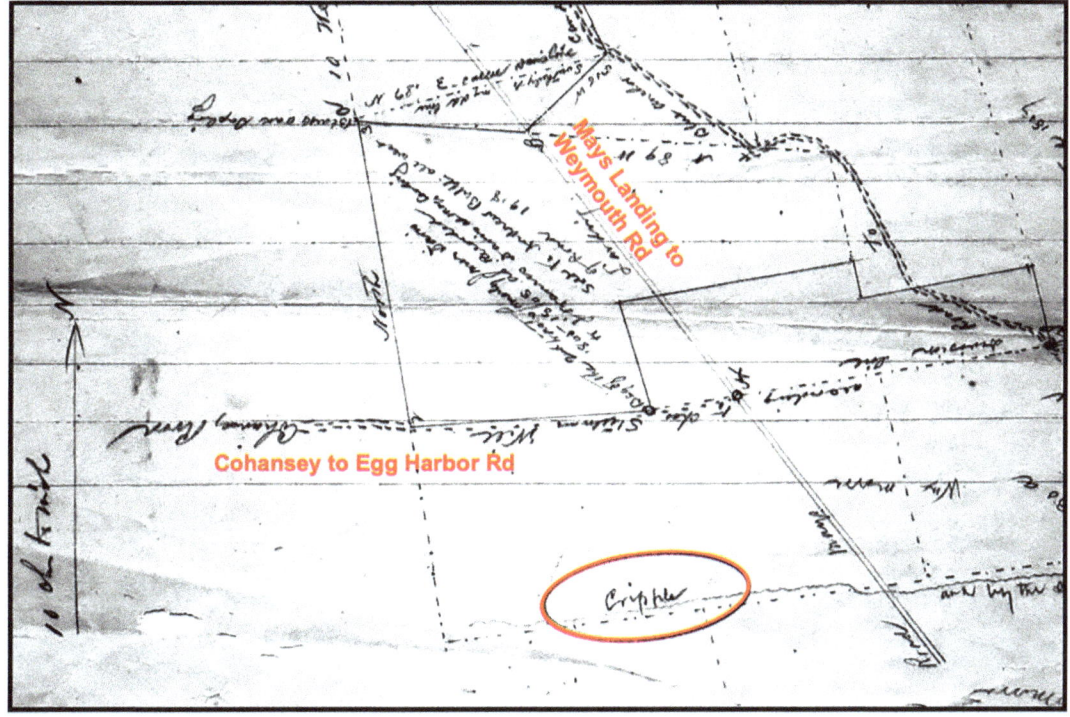

Fig. 5.4 Early map identifying cripple location.

The term "cripple" was commonly employed on surveys from the seventeenth to nineteenth centuries, but rarely employed in modern print. Cripples, as shown by Wright (1867b) and circled for emphasis by this author, are usually elongated to river-like wetland landforms, and rarely enclosed or spung-like (also see Figs. 5.1, 5.11). Emmelville, NJ.

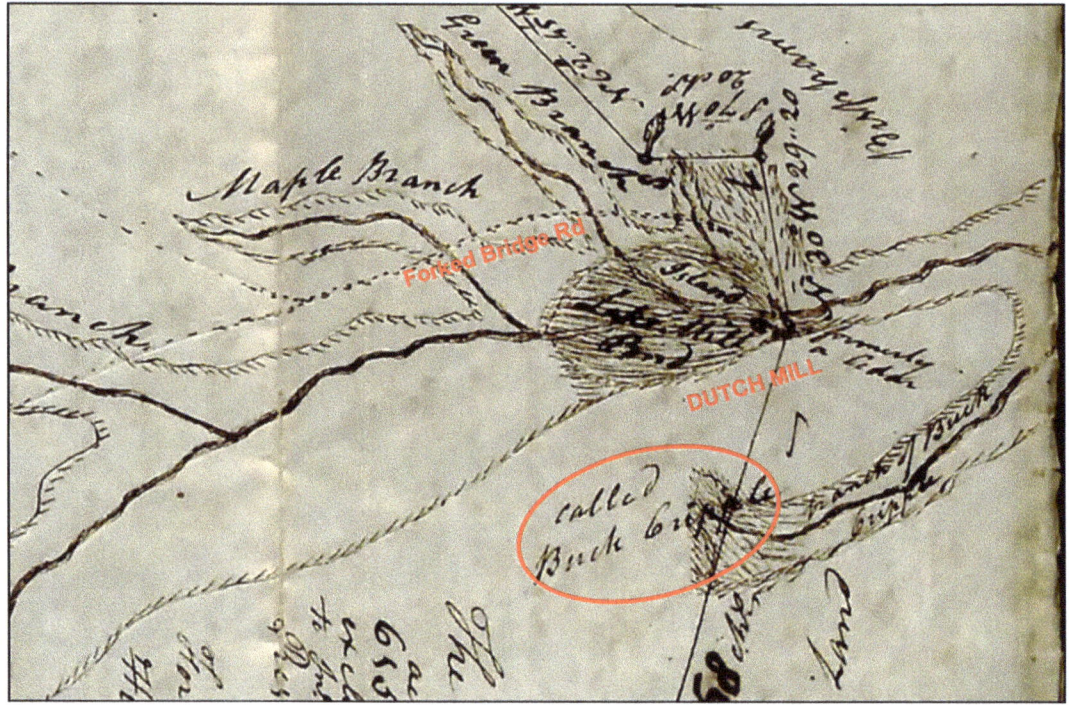

Fig. 5.5 Buck Cripple and a Branch of Buck Cripple on a map.

Clark & Hitchnor (1802), in resurvey of their Lake Tract, identified Buck Cripple. The Tract is named for a spung, the head pond on Lake Branch (Fig. 4.5C). Lake Mill, also called Millers Mill, later becomes Dutch, then Cedar Lake. SW of Collings Lakes, NJ.

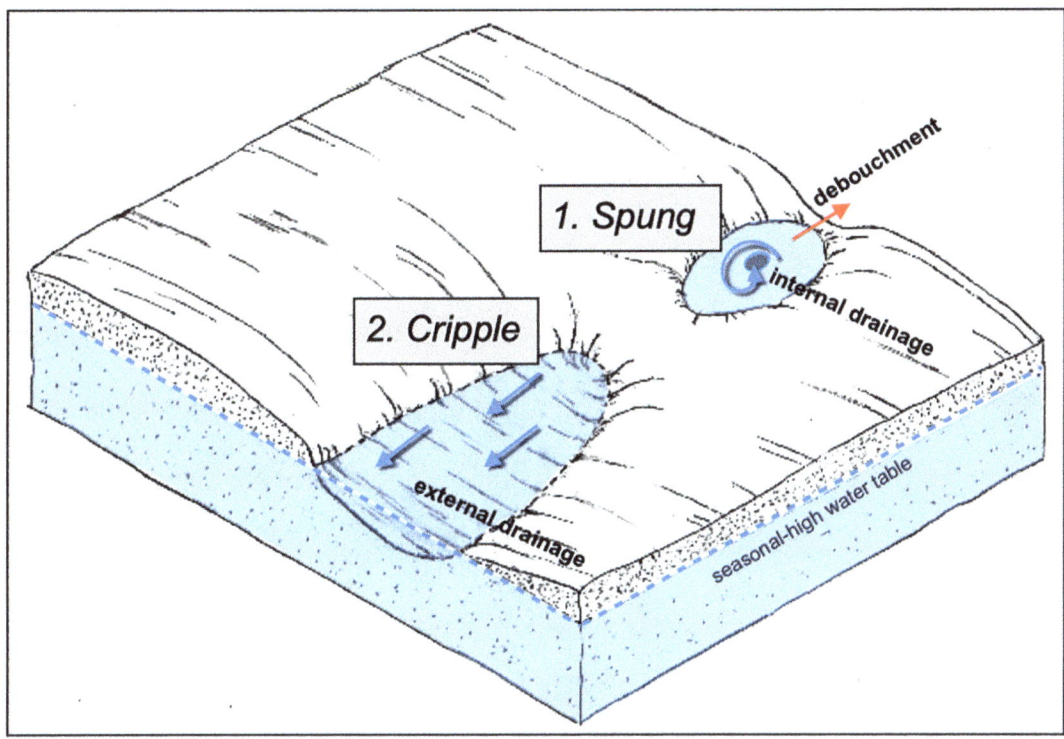

Fig. 5.6 Diagram illustrating water discharge differences between a spung and a cripple.

(1) Most spungs are closed systems where groundwater travels toward a central or "centripetal" (Wolfe 1953: 134, 1956: 508) drain. Bellybutton-like depressions are sometimes found in spung bottoms (Cresson 2000; Hajna 2002: 1A, photograph; Moore 2002a: B1, photograph), and may be piping relicts that formed when groundwater lowered during periods of aridity. Pipes are common to playa basins in arid regions today (Parker et al. 1990: 101–107; Gustavson et al. 1995; Castañeda et al. 2017). Under high infiltration phases, water can debouch a spung's rim and flow overland. (2) Cripples are open systems, and ephemeral sheet flow may occur across abandoned paleochannels under high infiltration phases. Within a short distance, cripples empty into larger drainage systems, as do periglacial dells in Europe (Dylik 1952; Troll 1957; Czudek 1986). Drawing by Pat Demitroff.

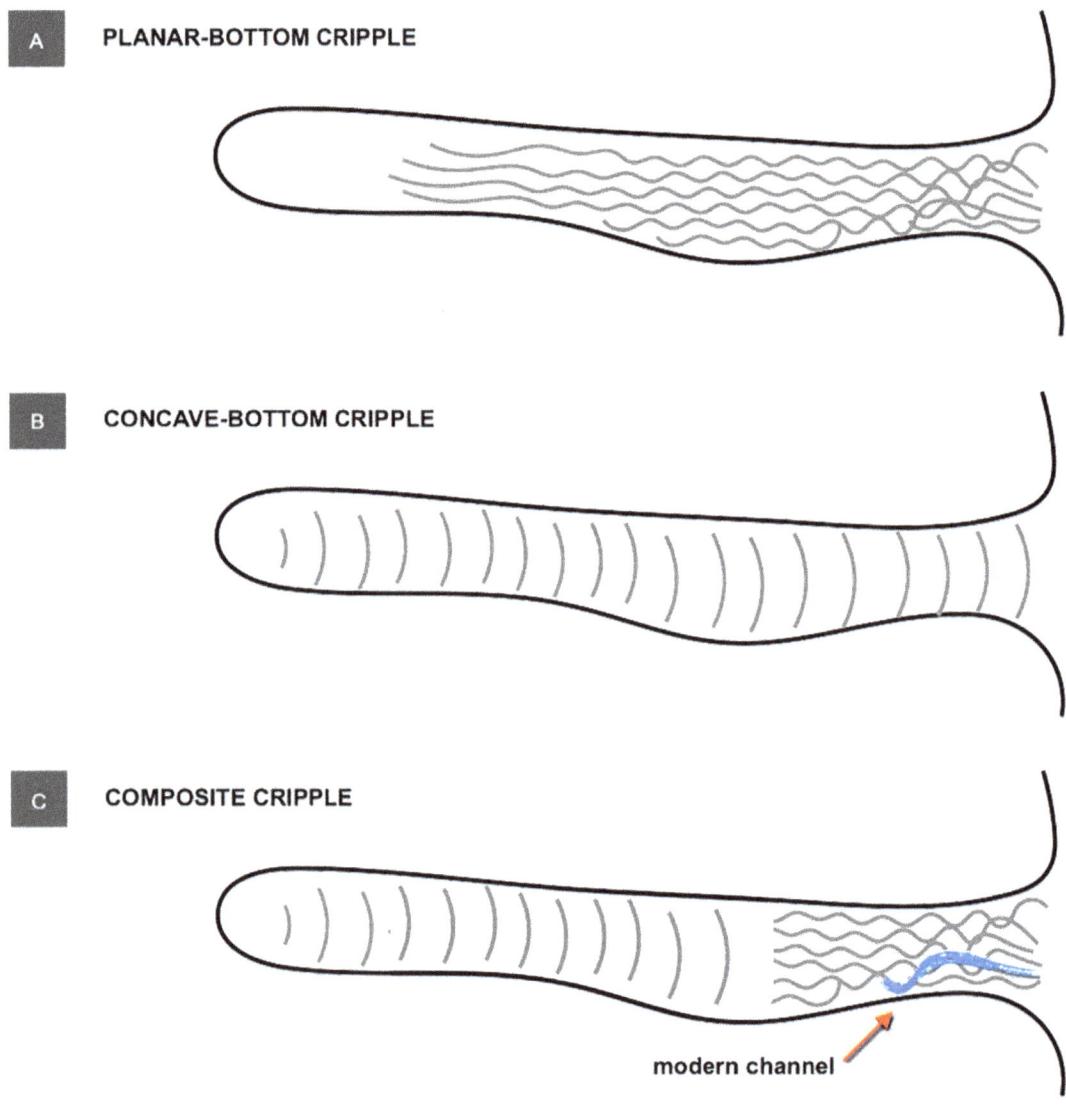

Fig. 5.7 List of cripple forms observed in South Jersey.

(A) Planar-bottom forms are often broad, flat, and complanate (are even with) with larger Late Pleistocene paleochannels. Fluvial wash over frozen ground was an important geomorphic agent.
(B) Planar-bottom forms can have narrow or broad channels, can be perched slightly above or complanate with larger Late Pleistocene paleochannels. Longer forms can have straight to curvilinear channels that vary widely in width along the length of the drainageway. Bank creep during permafrost thaw accompanied by wind-blown and sheet-washed sand were important geomorphic agents.
C) Composite channels begin as concave-bottomed channels that grade into planar-bottom channels before they discharge into larger Late Pleistocene paleochannels. With

A–C, deeper modern fluvial incision (red arrow) marks the cripples' conversion into an interglacial branch, run, brook, etc.

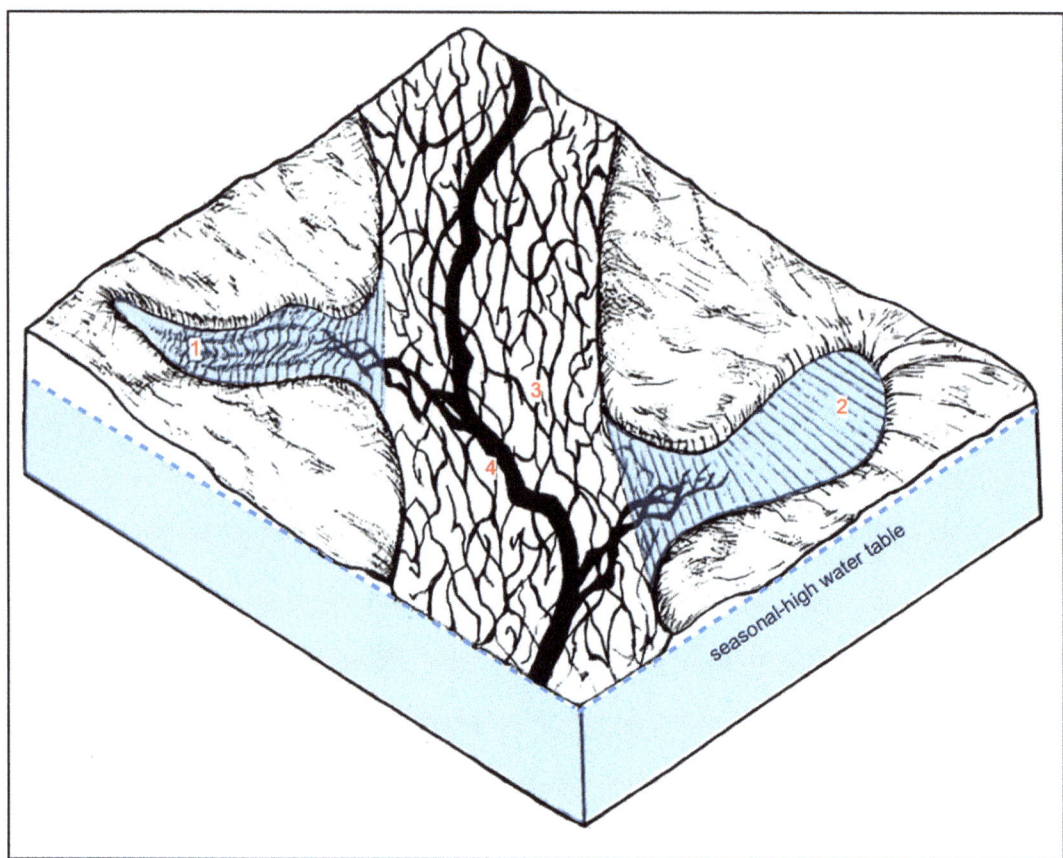

Fig. 5.8 Diagrammatic illustration of topographical relationships of generalized cripple types.

Both V-headed convex-bottomed forms (1) and round-headed planar-bottomed forms (2) have cripple floors that are complanate—are even with—larger Late Pleistocene paleochannels (3). With the disappearance of frozen ground, modern interglacial streams (4) incise deeper into the underlying sediments of the earlier paleochannel. The modern river (4) is misfit (underfit) to its bordering broad paleochannel (3). Drawing by Pat Demitroff.

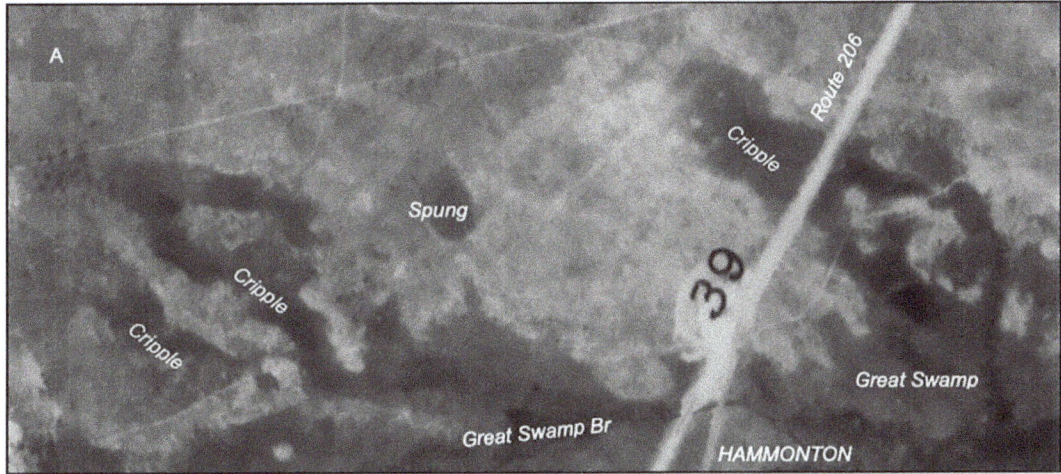

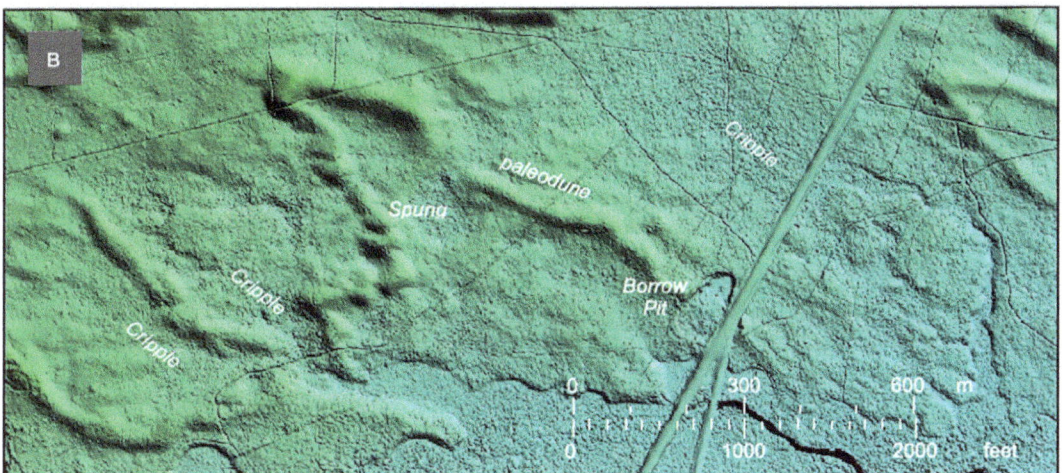

Fig. 5.9 Geographic setting of three planar-bottom cripples, Hammonton, NJ.

All drain onto Great Swamp Branch terraces by the edge of the West Mills dune field (Demitroff et al. 2019; Wolfe et al. 2023).

(A) The cripples appear as designated dark patches on early aerial photos. Aerial imagery (NJOIT-OGIS c. 1931, accessed through 1930s Aerial Photography of New Jersey, Boyd Ostroff).
(B) These cripples have deflated and sheet-washed to the level of an underlying Pleistocene alluvial terrace. Note how the spung is cripple-like but remains isolated by the deposition of a lunette on its southeastern rim. Lidar imagery (USGS & NJDEP, accessed as Lidar in the Pines 2022, Ostroff 2021).

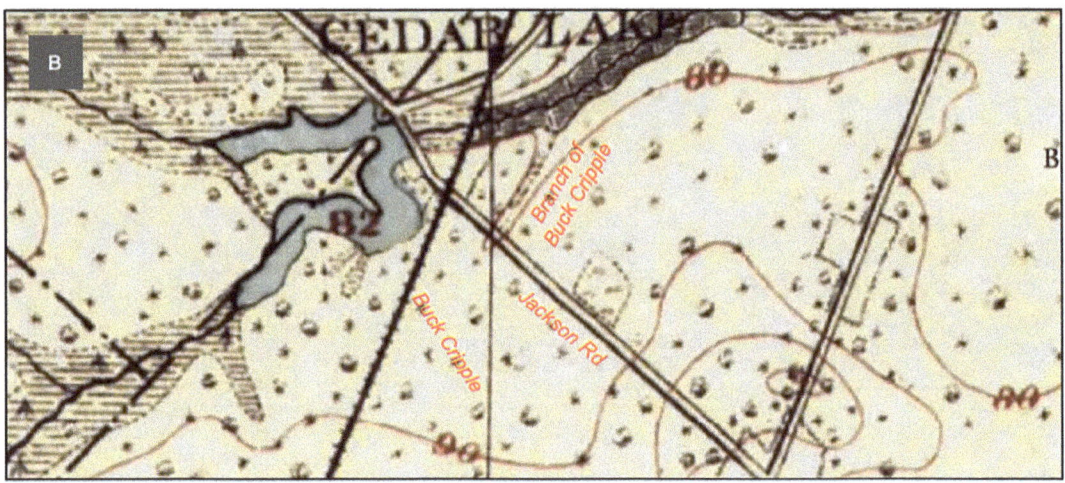

Fig. 5.10 Geographic setting of a planar-bottom cripple, Cedar Lake, NJ.

(A) The wetlands of Buck Cripple and a Branch of Buck Cripple (Clark & Hitchnan 1802) appear as dark patches on early aerial photos. Aerial imagery (NJOIT-OGIS c. 1931, accessed through 1930s Aerial Photography of New Jersey, Boyd Ostroff).
(B) Buck Cripple is not apparent on this early topographic sheet but the Branch of Buck Cripple does appear to be marked as a cripple-like valley. Resource extraction has since obliterated its trace. This area is the edge of the Newtonville dune field (Demitroff 2007; Wolfe et al. 2023). Map imagery (Cook & Vermeule 1890, Rutgers University, accessed as Cook's Map of the Pines, Boyd Ostroff 2019).

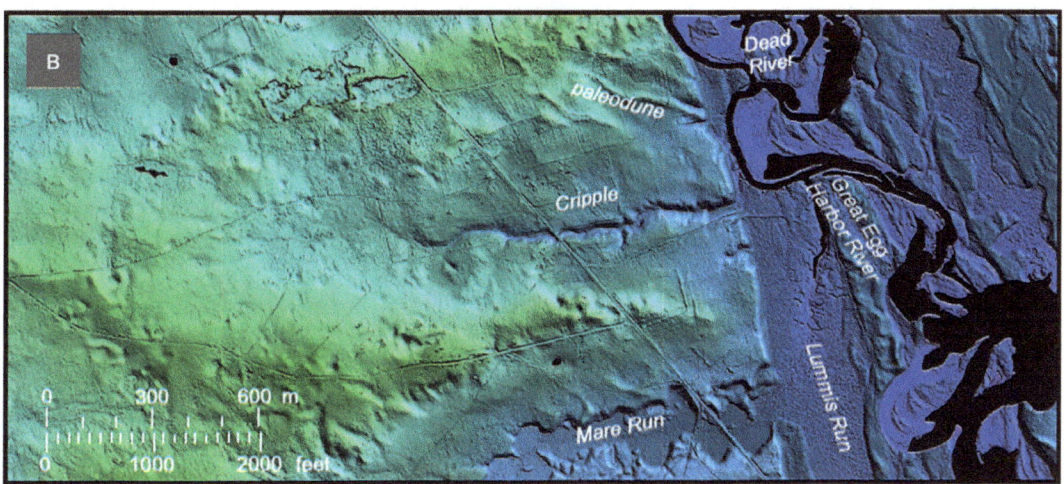

Fig. 5.11 Geographic setting of a concave-bottom cripple, Emmelville, NJ.

(A) In an early survey (Wright 1867b) this intermittent waterway north of Mare Run is said to be a "cripple." It appears as a short valley on Cook & Vermeule (1890 Rutgers University, accessed as Cook's Map of the Pines, Boyd Ostroff 2019).
(B) The cripple's extant can also be clearly seen on bare-earth laser altimetry. The cripple resides near the Lochs-of-the-Swamp dune field. An abundance of blowing sand accounts for the U-shaped valley of this cripple and adjacent paleodunes. Lidar imagery (USGS & NJDEP, accessed as Lidar in the Pines 2022, Ostroff 2021).

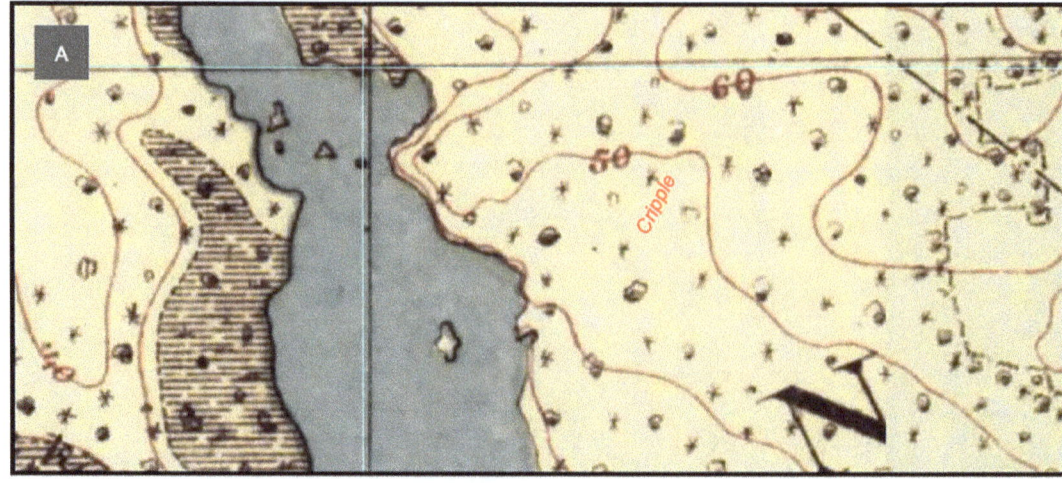

Fig. 5.12 Geographic setting of a concave-bottom cripple, Millville, NJ.

(A) Horse Wallow, later Horse Hollow, is a drainage corridor that passes through a Late Pleistocene dune complex (Newell et al. 2000) just below the Union House. As with many cripples, this wetland feature escapes a map record. Map imagery (Cook & Vermeule 1890, Rutgers University, accessed as Cook's Map of the Pines, Boyd Ostroff 2019).

(B) This watercourse runs perpendicular to the paleodune arms, the U-shaped channel being a byproduct of sand avulsion. Union is the site of an important ford, where a gravel bar (relict plateau) shoaled the Maurice River now flooded by Union Lake. Numerous ancient trails met here at a spring. The Union House was built in 1708 on the Malaga, Hance Bridge, and Old Cape trails. Lidar imagery (USGS & NJDEP, accessed as Lidar in the Pines 2022, Ostroff 2021).

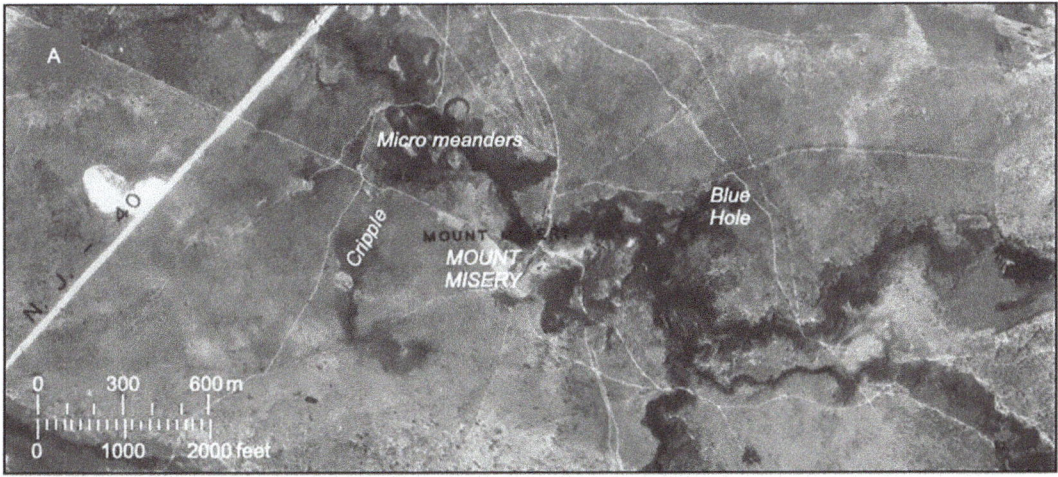

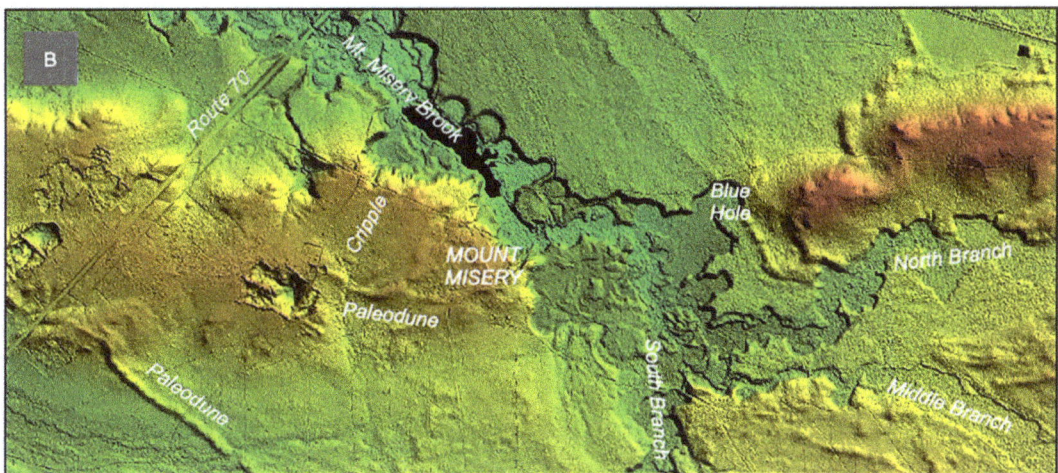

Fig. 5.13 Geographic setting of a concave-bottom cripple, Mount Misery, NJ.

The hamlet's oldest residents called this feature a "cripple" (personal communication, Gary Patterson, July 2003). The Mount Misery Blue Hole east of the cripple is addressed in Chapter 6.

(A) The cripple appears as dark patch on early aerial photos, its floor having deflated and sheet-washed to the level of an underlying Pleistocene meander terrace. Lidar imagery (NJOIT-OGIS c. 1931, accessed through 1930s Aerial Photography of New Jersey, Boyd Ostroff).
(B) Note how a narrow channel has incised into a dune dam between the words "Cripple" and "Paleodune" on the above imagery. Lidar imagery (USGS & NJDEP, accessed as Lidar in the Pines 2022, Ostroff 2021).

06
Blue Holes of the Pine Barrens

Etymology of Blue Holes

As with cripples, "blue holes" of the Pine Barrens are an all but unknown landform (Figs. 6.1–6.2) (cf. Mylroie et al. 1995, karstic blue holes). They are often deep, circular cavities occupying the bottom of streambeds. Strong riverine upwellings or "boiling springs" issue—at least in the recent past—sparkling blue water under a strong hydrostatic head. Their "turquoise blue" waters contrast with the typical tea-colored "cedar waters" of the Pinelands water bodies. Deep places in rivers have long been known as "holes" (OED 1899).

In Pine Barrens folklore, each hole is as sinister as the next. All are reputed to be bottomless, and all possess dangerous "whirlpools." Young bathers were duly warned not to swim over them, as the icy jets of water emanating from their fathomless depths would cause immediate cramping. You could sink like a stone in the brisk current, or worse yet, be pulled under by the "suckhole."[1] Or was it really the fault of the Jersey Devil, abroad from his subterranean lair, hunting for yet another victim?

The first to associate the eerie bluish cast of Inskeep's Blue Hole with a particular species of algae was Pennypacker (1936: 93). The Mount Misery Blue Hole may receive its namesake from the hue of the blue-green algae

---

1. It is surmised that the strong spring that emanated here produced quicksand (Neuendorf et al. 2005: 531, "A mass or bed of fine sand . . . that consists of smooth rounded grains with little tendency to mutual adherence and that is usually saturated with water flowing upward through the voids, forming a soft, shifting, semiliquid, highly mobile mass that yields easily to pressure and hence is apt to engulf persons and animals coming upon it."

present there too (personal communications, Walter Bien, July 2004). On a visit to the Inskeeps Blue Hole on August 18, 2004, this author and Bien confirmed the presence of a similar alga or complex of algae discovered at the Mount Misery Blue Hole (Fig. 6.3). The algal bloom at Inskeeps may be related to reactivation of springs by recent downpours. This author remembers a similar bluish cast to the upwelling at Inskeeps Ford many years ago. It is possible that the nitrogen-poor waters issuing from deep upwellings are somehow conducive to microfloral growth. Boyd (2008: 32) suggested that blue holes contain massive growths of *Cyanophyta*. Other strong springs with colored waters have been found to possess interesting vascular plant (Mowszowicz & Olaczek 1961; Pinkava 1963), microbial (Schwabe & Herbert 2004), diatom (Rakowska 1996), and insect (Melin & Graves 1971) communities. Makhalanyane et al. (2015: 827) reports very high levels of microbial activity even in the Antarctic Dry Valleys, as represented by heterocytous and non-heterocytous filamentous mats.

An unusual diatom species was discovered at Inskeeps by Charles Boyer[2] of the Philadelphia Academy of Natural Sciences in the 1920s and published by Reimer (1959; Hamilton et al. 2004) as *Neidium rudimentarum*. The genus is primarily found in cold climates. As a species this diatom has only been found in two places, Inskeeps Blue Hole and Crane Pond in New Hampshire. Its trace has not been found in South Jersey in recent times and it may be fossil species (personal communication, Paul Hamilton, July 07, 2020). This author suspects that modern sediments may have since buried the older fossil-bearing deposits as this water feature shallowed in recent years.

Stories of Blue Holes

As a young boy in the Pine Barrens, the author heard many wondrous tales of these bottomless pools scattered about the various riverbeds. Kathryn Chalmers,[3] from New Germany (today Folsom), gave the earliest

---

2. Charles F. Boyer (1856-1924), a school administrator and prominent diatomist (Reimer et al. 1991).

3. Kathryn H. Chalmers (1879-1947) was a student of English at Columbia and Harvard Universities. She envisioned the Long-a-Coming trail to be preserved as a linear living museum of history and nature (Eckhardt 1975: 109), not unlike the Skyline Drive in Shenandoah National Park (personal communication, Frances Kathleen Corsiglia Zambone, 1999). Schoolteacher Zambone (1901-2009) was born on a Richland farm adjacent to this author, student taught c. 1922 under Chalmers at Folsom School, and

account of the most celebrated of these upwellings, called the "Inskeeps Blue Hole."[4] Chalmers had lifelong knowledge of the feature. She had been raised on the Long-a-Coming trail in New Germany, just downstream from the Inskeeps Blue Hole. She described the mysterious blue hole as unswimmable due to its icy springs (Chalmers 1951: 84–86). To the schoolteacher, the water's

> Marvelous transparency reflecting heaven's blue," and no matter how cold it became, "portions . . . never freeze over but send up a continuous fog-like steam in the coldest weather. The [ditched] outlets have become clogged so that the water moves slowly toward the center where the funnel-shaped bottom forms a whirlpool. One story has it that the water of the whirlpool drains downward into the Atlantic Ocean.

Chalmers then adds, "frequently have scientists tried to measure the depth of this center vortex but their lines were never long enough to reach the bottom." Creek dredging by federal workers to draw off "the stagnant water" raised concerns that water drainage would become too rapid and desiccate the surrounding Barrens (Chalmers 1951: 140). Another early account is provided by Pennypacker[5] (1936: 90–96), which includes folklore and local history.

Inskeeps' legendary pool received popular attention with a publication by Henry Charlton Beck (1937), a compilation of historical vignettes written as articles in the *Courier Post*. Beck's account is full of the lore told over and over in the Pines: the blue hole is bottomless, there is a whirlpool, the ice-cold water is tinted blue, and this devil's hole was the site of paranormal happenings recounted in many tales. In a photograph caption, Beck (1945: 240–250) revealed that the Jersey Devil lurks within "darksome pools" of the Pine Barrens (Beck 1945, opposite page 21). The enclosed

---

shared many fascinating accounts of local lore with this author over her many years.

4. Locally the site is often called Inskips or Inkskips—not Inskeeps—Blue Hole. Apparently, the surname was elastic (e.g., Inskip for Inskeep — Mattocks 1762; White 1785: 38).

5. James Lane Pennypacker (1855–1934), resident of Haddonfield, was a noted publisher, writer, poet, and member of the Academy of Natural Sciences of Philadelphia. Pennypacker Park in Haddonfield, where the first nearly complete dinosaur *Hadrosaurus foulkii* was found in 1858, is his namesake (Levins 2008). Coincidently, Chalmers taught in Haddonfield, and included a chapter on the community in her book—being on the Long-a-Coming trail.

basins and depressions forming such pools are known as "spungs" and "blue holes" by the locals. Sinister stories about the blue hole are retold in various newspaper, book, and magazine articles (Ribeiro 1996a: 6–8, 1996b: 42–43; Leap[6] 1996: 45–47; McDonald 1999; Bennett 1999; Walsh 2003; Sceurman & Moran 2004: 43–45; Polhamas 2015), and at various sites on the Internet.

Two written references cite other blue holes. In Haddonfield, a deep riverine pit appears at the head of a short branch. This small stream quickly drains into the Cooper River. The deep hole first appeared on a 1686 map (Farr 2002) and is named Dicks Hole. Another is "Dog Heaven" (Anderson 1964: 68), near an old charcoal camp in Pleasantville by Charcoal Hill. It too had fathomless depth, whirlpools, and a dastardly devil thrown in for good measure. It was said "When men and their dogs were out hunting for the Devil and came near to this whirlpool, the dogs dropped dead." The bottomless nature of these Pinelands pools is by no means unique in the geographical lore. This folkloric landscape of hazardous pools is shared around the world (Meredith 2002; Gladden 2011), even in Siberia (Woo 2012: 107, Fig. 3.30).

Blue Hole Morphology

Blue holes (Figs. 6.4–6.5) usually occur within modern (Holocene) stream channels. As a boy, this author would wade and swim the tea-stained waters of various creeks and branches with his schoolmates. Periodically, springs were encountered along the gravelly bottoms of Pine Barrens watercourses. Their presence was duly noted, as the normally lukewarm waters would turn icy-cold upon passing over them. The largest springs were associated with deep cavities in the riverbed (cf. Rose et al. 1878: 21, springs make deep holes in marsh creeks), making excellent swimming holes. Riverine shallow spung-like ponds were sometimes found just downstream from the larger upwellings. Some ponded watercourses lack a distinct springhead and are likely associated with groundwater seep through attendant ancient gravel bar deposits.

The largest of the blue holes, Inskeeps (Figs. 6.1, 6.6, 6.12A) and Mount Misery (Figs. 6.2, 6.7), do not reside within a modern riverbed. They instead

---

6. Historian William Leap proposed to his spouse, conservationist Clare F. Leap, in 1953 at the Inskeeps Blue Hole. Comparably, this author proposed to his wife in 1988 at Success—site of an old cranberry bog—by the edge of a spung.

occupy ancient broad paleochannels bordering smaller, modern underfit stream channels. Both upwellings are present where marked changes in paleochannel morphology occur in association with gravel bar caps just upstream from each blue hole (Figs. 6.8–6.9). Ancient trails follow these gravel deposits across both rivers, as these sites are opportune locations for fording. Here stream-corridor bottoms are shoaled with gravel pavement facilitating wheeled traffic passage.

Early Blue Hole Use and Topography

Like spungs, blue holes often had precontact trails that were contiguous with their shores (Fig. 6.10). The Long-a-Coming trail passes just to the north of the Inskeeps Blue Hole and meets with The Blue Anchor (Boyer 1962: 64) and Shamong (Philhower 1931; Stewart 1932) trails at this spot. Pennypacker (1936: 93) notes this place as an intersection of well-used Indian trails—long known as the Blue Hole by old inhabitants for miles around. This trail juncture is the site of long aboriginal occupation (Blackman 1880: 404) from the Archaic through the Woodland times (Mounier 1972), although Mounier suspects that the site is even older (personal communications, April 2004). A trail crossed Stony Brook at Dog Heaven (Anderson 1964: 68). In another case multiple trails converged at the Davenport Hole (Mizpah) on South River.

The presence of a proximal fresh water source (e.g., spring, pond, stream) is important for prehistoric Indigenous utilization of a site (Haynes & Agogino 1966; Berón 2016). Raised ancient gravel bar cuts, when eroded into by river currents, expose seeps and springs. Although precipitation is plentiful in the Pines, the permeable soils allow rainwater to infiltrate quickly, creating desert-like edaphic conditions in the uplands. Strong springs once flowed in scattered lowland locations, and not all of them were known as "blue holes." Pamphylia Spring[7] in Bridgeton (Elmer 1869: 30; Cushing & Sheppard 1883: 505–507, 515) is a good example of a strong riverine upwelling with characteristics of a blue hole (Cresson et al. 2006,

---

7. Coincidently, Mounier (1972: 26) used the closed component assemblages of Skinner and Schrabisch (1913) at Pamphylia Spring as an analog to the Inskeeps Blue Hole since both sites contained a single component or multiple components of the same cultural period. Because both sites were situated at perpetual springs, a greater range of archaeological cultures would be expected (personal communication, Alan Mounier, September 17, 2006). Pamphylia is ancient Greek for "land of all tribes," and was considered a colorful place where many factions of Greeks met (Magie 2016: 261).

i.e., precontact settlement, converging trails, a river ford provided by a gravel[8] outcropping), yet there is no record of it being called a blue hole. Mesolithic artifacts were found in association with hollows interpreted as the sites of fossil spring sites called icings (Catt et al. 1982—Wessex Downland). Perhaps blue holes were as mysterious and enigmatic to aboriginal people as they were to those of the historic period. Nearby prehistoric settlements are not well understood and warrant further investigation.

Other blue holes existed but have not found their way into the literature. The Mount Misery Blue Hole was the subject of a doctoral dissertation comparing the ecology of two species of *Sphagnum* (Bien 1999). It is discernible as a distinct blue dot on the US Geological Survey, Browns Mills Quadrangle map[9] (Fig. 6.7).

In Newtonville, New Jersey, yet another blue hole (Bennett 1999) in an account by storyteller Michelle Washington Wilson was the site of a sad tale about a large "suckhole" that mysteriously appeared and claimed a victim. Blacks were prohibited from swimming and baptizing at nearby Lake Lenape at the time of the mishap and used the blue hole in its stead. The modern Newtonville Blue Hole is actually a manmade extraction pit for the nearby NJ State Route 54 bridge embankment and borrows its name from a natural blue hole about 1 kilometer (0.6 miles) to the south on Money Run, today misnamed Three Pond Branch (Clark & Hitchnor 1802, termed an alias Three Pond Branch). The natural blue hole was above the Jackson Road bridge and was reputed to be as sinister as the other holes. "Put anything in it, and you would never see it again" (personal communication, Annie Huff & Peggy Jones, 2000). Another claimed blue hole was located just south of the Unexpected Road bridge, less than one km (half mile) to the south of the Jackson Road blue hole, again on Money Run.

Along the South River at the Main Avenue bridge in Richland, two old metal signs forewarned bathers with their message of "Danger Deep

---

8. Newell et al. (2000) noted the widespread presence of ancient gravel bars associated with the proto-Hudson's passage over the Outer Coastal Plain (see Demitroff 2016: 125). Pebbles and cobbles associated with blue holes provided a source of otherwise scarce lithic material for precontact tool use in otherwise sandy rock-poor terrain (e.g., hammers, points, boiling stones). Gravel bars (and ironstone capping) can form choke points in river corridors, helping to create sturdy substrate for mill and dam building.

9. Christian Bethmann (superintendent of the Brendan Byrne State Forest), Gary Patterson (Rowan University) and the then Pinelands Commissioner, Ted Gordon, revealed the existence of this spring to me.

Water." This location, the "Danger Hole" (Figs. 6.8, 6.12B, 8.8), was a favorite resting spot midway between the author's boyhood farm and his father's feed mill. An old schoolteacher of the author (personal communication, Frances Zambone, 1975) recounts that two boys drowned at this spot by diving from an old ironstone bridge, which was subsequently washed out by a storm in the early 1980s. Three km (1.8 miles) downriver, another hole was at the Davenport Place. Ancient trails converged there, and it was a favorite meeting place for "coalers" (charcoal makers) to bathe. When Russian Jews settled nearby at Mizpah in 1891 (Marsh et al. 2019), they established a religious bathhouse or "mikveh" at this location (personal communication, Virginia Davenport Gale, February 2005).

The Zion Meeting House in Bargaintown was established c. 1764 (Willis et al. 1915: 101–102) at the Cedar Bridge on a spur of the Long-a-Coming trail beginning at Gravelly Run. In local lore, a riverine pond in the small branch behind the church is bottomless, and it shares the lore of other blue holes (personal communication, Fotios Tjoumakaris, January 2003). The neighboring farmer (personal communication, John Prantl, 2003) modified this hole when its springs no longer produced enough irrigation water for his tomato field.

Origin of Blue Holes

The academic community did not recognize blue holes until the GANJ conference at Richard Stockton College[10] (Demitroff 2003). They remained in the realm of Pinelands legend and folklore; mythical locales replete with fantastic stories. Several possible origins are offered by Chalmers: 1) a marl-bed—perhaps referring to a sink formed by the solution of carbonates; 2) a meteorite strike; and 3) quicksand (Chalmers 1951: 97, 122–124). A short paper by Powell (1958) considered the problem of the Inskeeps Blue Hole from a geologic perspective. She made some observations, tested the waters, sampled the surrounding soil, and sounded the pool's depths concluding: 1) the water entering the pool is not issuing from a great depth; 2) the blue tint is from sky reflection; and 3) the pool was likely created by excavation of material for a nearby dam's earthworks.[11]

---

10. Now Stockton University.

11. It is important to note that Inskeeps dam (Farr 2018, c. 1765) was built at a time that predated a technology to dig below the water table. Instead, this author noted a series

A newspaper article by Hilton (1999) describes how Millikin, a geologist from the New Jersey Geological Survey, was invited to search for the legendary blue hole. After some site investigation, he was unable to explain how the hole got there. Millikin cautiously stated, "No matter what your belief in the folklore or the Jersey Devil [a legendary demon], or even the practical explanation of the [erosive activity of a] mill—it's a great mystery" (Hilton 1999: 3). In another newspaper article geologist Gail Ashley suggests Inskeep was once an ice mound site known as a "pingo" (Bennett 1999).

Some insight into the nature of the Inskeeps Blue Hole can be obtained from c. 1931 aerial photomosaics of the area (Fig. 6.11). From its source in Berlin,[12] and on to the Inskeeps Blue Hole, the Great Egg Harbor River occupies a broad channel bottom, a channel typical of braided rivers in the High Arctic today (Hamelin & Cook 1967: 115; van Vliet-Lanoë 2005: 378, Fig. 15.8; French & Demitroff 2012: 33), a channel-pattern phenomenon first reported in the Pinelands by Newell and Wyckoff (1992). In both Pleistocene and modern context, many braided periglacial streams formed during the presence of widespread permafrost (Vandenberghe & Woo 2002: 495). Polar streams in small basins are usually ephemeral, with nival floods in late spring and little flow by late summer (Woo 1993; French 1996: 192, Fig. 11.5; Woo & Pomeroy 2012: 118–120). In cold semiarid regions, 30–90% of the annual runoff occurs within a few days or weeks (Vandenberghe & Woo 2002). Where poorly vegetated, as was probably the case in the Pinelands during the cold period of the Late Pleistocene, the streams deposited their large suspended load during this short flood period, further encouraging the braided pattern. Catchment sources for Pine Barrens rivers were composed of unconsolidated sands, which as periglacial sandur (Ballantyne 2018: 268) readily washed into the braided paleochannels during the brief active snowmelt stage (van Amerom 1990; Kasse 1998: 93, 94). Eolian and fluvio-eolian processes were also important, but episodic, sediment sources to local waterways (Schokker 2003; French & Demitroff 2012).

---

of shallow scour channels that radiate to the southeast of the dam, indicating the embankment fill had been scooped out in this manner (see Farr 2018, noted during "A small group of local historians, including this author, visited the site in March of 2000).

12. The Great Egg Harbor River headed at a spung called Long-a-Coming (Chalmers 1951: 46, photo of source opposite page), from the Algonquin word Lonaconin(g) meaning "place where waters meet" that originated north of Franklin Avenue in Berlin Township (Farr 2002: 65).

Below the Inskeeps Blue Hole, the river channel looks very different. A series of meanders or cuspated channels appear. Meanders in cold-based rivers are often related to transient, modest energy fluvial systems (Vandenberghe 2001: 114; & Sidorchuk 2020), in contrast with high-energy braided fluvial systems. Such unusual channels could be explained as an artifact of the presence of a perennial spring at the present blue hole site when a cold climate was present, and the after-effects associated with icing ablation.

Icing Building

The spring itself would provide the water needed to maintain a year-round pool of unfrozen water charge. Such springs exist in the Arctic today and are known to cause "groundwater" (Sokolov et al. 1989: 240–243) or "spring" (Åkerman 1982) icings. They are "common, but neglected features" (Åkerman 1982: 190) and "cover large areas of the floodplains of braided streams" (Åkerman 1982: 192; also see Mikhailov 2016: 43) with thick tabular masses of ice known as *aufeis* following the convention of Ensom et al. (2020). They can be compared with glaciers in ice volume and action, although their modes of genesis differ (Sheinkman 2012; 2016). Icings are ubiquitous in polar regions today (Sokolov et al. 1989: 233; Alekseyev 1998: 211), as they were during Pleistocene periods of "cryoaridization" across Siberia (Sheinkman 2012, 2016). Furthermore, they were almost certainly present in smaller size but higher number (Ensom et al. 2020) considering the gravelly sandy Late Pleistocene planar landscape of southern New Jersey. During the glacial epoch the Pinelands experienced a severe, continental climate with minimal snowfall (French & Demitroff 2003, et al. 2003, semi-desert; Demitroff 2016, desert). Such conditions would have been conducive to the development of distinctive icings (Yershov 1998: 211–212; Alekseyev 2015; Ensom et al. 2020). More importantly, icings would have increased as the climate moderated at the end of the Wisconsinan (Michel 1994; Morse & Wolfe 2017), so their traces might be easier to discern, as they are relicts of permafrost degradation to deep seasonal frost (i.e., younger) rather than permafrost aggradation (i.e., older).

The blue hole at Mount Misery, like that at Inskeeps, issues from a broad paleochannel adjacent to a stream. Its appearance also demarcates the transition from a broad ephemeral paleochannel to a cuspated paleochannel along the North Branch of the Mount Misery Brook. Both the Mount Misery and Inskeeps springs occur where a waterway crosses a relict plateau divide capped by sand, gravel, and clay (Figs. 6.6–6.7).

Strong springs are present at both sites. Both sites reside west of their relict plateau, where the nearby plateau slope would have received the maximum solar insolation to help melt frozen groundwater to charge the springs. It is reasonable to assume that the permeable strata and topography responsible for today's springhead was in place for a considerable time. The sites of contemporary upwellings are likely to have been the locations of ancient icings too (hydrological memory). In the Canadian Arctic, icing sites are associated with granular deposits that allow the passage of groundwater through taliks in permafrost (Veillette & Thomas 1979: 795; Ensom et al. 2020). Old proto-Hudson gravel bars are commonplace (Newell et al. 2000), places where—under frozen ground conditions—gravel lenses would have remained partially saturated and unfrozen, becoming conduits for the groundwater seep required for icing genesis (Woo 2012: 73–118). The infiltration and exfiltration of groundwater in the paleochannel leads to a significant heat exchange between the river and the gravel bars, providing significant thaw zones (taliks) subject to complex diurnal and seasonal changes (Ensom et al. 2020; Mikhailov 1998, NE Siberia; Alekseyev 2015, NE Siberia). Humlum (1979) reported the preservation of icing-site remains in modern Greenland in association with braided channels.

Icing Degradation

It is possible that these springs were active during the cold periods of the Late Pleistocene, forming riverine pools below them (Fig. 6.12). In example, the South River and Money Run each have broad riverine ponds just downstream from their reputed blue holes. Boreham (1997) described a tear-shaped riverine hollow in Essex, UK, as an aufeis relict. Closed depressions found within dry flat-floored periglacial braided channels are attributed to the melting of buried aufeis in northern Poland (Kozarski 1975; 1995: 89). A shallow spring-fed basin in East Anglia, Buggs Hole (Coxon 1978), is presumed to be an aufeis relict. Catt et al. (1982) interpreted a series of mounds and hollows as aufeis remnants, and van Vliet-Lanoë et al. (2017) noted aufeis relicts in the St. Aube River valley sands and gravels of northern France. Although Worsley (1985; 1997) found it difficult to deny the possibility of fossil aufeis structures, strong evidence for their presence was lacking and their preservation potential would be very low.

In wide flat channels, icing occurrences produce extensive tabular masses (Williams & Smith 1989: 207) that substantially influence their surroundings (Alekseyev 1998: 217; 2015). Icings and their aufeis share many characteristics with glaciers and can even be placed in the glacier geosystem complex" (Sokolov et al. 1989: 243–244). It is likely that during the coldest episodes the ice-marginal Pine Barrens aufeis would become perennial in nature and even incorporate with long-term snow patches to form small glacier-like structures.

Mikhailov (1998: 336) noted in Siberia that icing-forming taliks generate below river damming,[13] which may help explain why icings might preferentially occur below gravel-bar/river intersections. Downstream from an ice-forming spring, the thickness of an icing can often transform the valley vegetation into "primitive groupings typical of polar deserts" (Sokolov et al. 1989: 243–244; also Alekseyev 1998: 219), and produce wide "icing clearings" (Sokolov et al. 1989: 242) or "icing glades" of fresh green meadow (Alekseyev 2015) that are subject to an enhanced weathering of the channel sediments and channel form widening by icing damming (Sheinkman 2012, 2016), which would in turn make the broadened glade bottoms subject to deflation (Alekseyev 2015: 187, Fig. 9). Newell and Wyckoff (1992: 25) noted windblown frosted sands and ventifact pavements[14] at the base of riverine dunes in the Pine Barrens. Such a

---

13. A detachment block has been recorded in the Pine Barrens sedimentary record, evidence of a Jökulhlaup-like ice-dam flooding event (Demitroff 2016: 132–133). Its extant records an occurrence of bank collapse and high turbulence flooding consistent with an ice-dam-break event.

14. Rock varnish or glaze often accompanied wind sculpting of stones in southern New Jersey. Tremblay (1961: 1561) reported that ventifact surfaces throughout Canada exhibit a high gloss. Glossy rock varnish was also noted in other polar regions (Seppälä 2004: 136–137).

Blowing dust has been suggested as the source of desert glaze (Livingstone & Warren 1996: 56–57) and rock varnish (Dorn et al. 2013). Thiagarajan and Lee (2004) ruled out the possibility of rock varnish formation through the deposition of dust particles alone since certain trace elemental anomalies could only be explained if the depositional processes occurred in an aqueous environment. They proposed that dust was entrained in rain or fog, and certain constituents dissolved into solution. According to Thiagarajan and Lee (2004), these mineral-laden droplets were later deposited on lithic surfaces and formed rock coatings.

But what if the dust was encapsulated in ice or snow? Dusty events accompanied certain cold and windy periods of the Pleistocene (Pye 1995; French 2018: 355–358). Ongoing work of Demitroff et al. (2019) provisionally indicates the Pine Barrens atmosphere was dust laden during or after hairpin parabolic dune building. Ice crystals alone have been shown to abrade and polish rock (Seppälä 2004: 130–132; Schylter 1994), although Bridges and Laity (2013: 144) found no evidence to support this. It is possible that under periglacial conditions, physical and physiochemical processes aid

mechanism could, for example, help account for the well-developed paleodunes along the cuspated channels below the Inskeeps Blue Hole, having formed downwind of NNW–SSE flowing katabatic winds, (Figs. 6.13–6.14). A relationship between eolian deposits, icings, and river meanders is suspected at locations throughout the Pines (e.g., Fig. 5.13, Mount Misery; Fig. 6.11, Inskeeps).[15] Most remarkable is the geomorphology near Whiting, where the interplay between wind action and paleohydrology is most confounding (Fig. 6.15).

Blue Holes: Summary

Many issues arise concerning the status of these Pinelands springs. Although springs generally drew attention in earlier works (Cook 1857: 129–131, *springs*; Elmer 1869: 30, 48–49 *Pamphylia Spring*; Rose et al. 1878: 21, *boiling springs*; Taylor 1880b: 3, *boiling spring*; Bowen 1885:

---

in the erosivity of ice crystals (Demitroff 2016: 130). The addition of dust particles to ice crystals would increase ice erosiveness. The entrainment of dust in a frozen form may add a new dimension to the work of Thiagarajan and Lee (2004)—cf. Siman-Tov et al. (2017, the accretion of glacial polish).

15. Mikhailov (2016) noted that river meanders in the frozen ground of Siberia occurred in association with tectonic block sinking along former broad-braided river channels, which resulted in a diminution in both channel width and gradient. Initially, the broad-braided channel acted as a talik, allowing heat and mass exchange across the entire channel. An erosional lowering of the riverbed resulted in a subsequent freezing of the previously unfrozen groundwater talik. The former braided riverbed in response becomes a frozen elevated terrace, disallowing heat and mass exchange across the entire channel. This then narrows the width of the stream channel, which results in over-bank flooding common to meandered channels, especially where sand bars and pebble layers have been overlain by sand and loamy sand (p. 39). It is a reduction of channel width that results in a proportional diminution in water surface area that was needed to facilitate heat transfer primarily through solar radiation.

Since the Pine Barrens is tectonically quiescent, it is suggested that—in lieu of tectonic block sinking—here river bottoms drop in response to climate amelioration and increasing thaw depth. Thus narrowed, a broad-braided channel will switch to a meandered one. Along with this climate warming, once-frozen sands would increasingly be available for wind entrainment. Stream channels are then further narrowed by sand avulsion, increasing the meandering effect due to channel narrowing. In thaw, renewed taliks and groundwater flow (both above a cryogenic aquiclude and just seep in general) could further promote stream meandering (see van Balen et al. 2008) in Europe's Sand Belt. Eekhout et al. (2013) like van Balen, suggested seep through dune sand as a cause of meandering in southeast Netherlands, and in adjacent Germany—both being past permafrost areas.

39, *boiling springs*; Beck 1937: 148–153, *blue hole*; Weiss & Kemble 1962, mineral springs), they are almost but not completely absent from the modern literature (Domber et al. 2015; NJDEP 2019). What happened to the numerous springs of the older works and spoken accounts? As with spungs and cripples, what was (and is) the extent of Pinelands springs? Where is the source of their water? Could the past presence of Pleistocene permafrost have left a distinctive mark on streams in the Pine Barrens? Are there floral and faunal signatures unique to these upwellings?

Strong springs, once commonplace in the Pine Barrens, are now a rarity. Stories of the blue holes' unusual properties give important clues as to the nature of the environment when these stories were constructed (Meredith 2002: 331). Too often, we ignore or dismiss the collective memory of a place. Regarding the conflict between folklore and science, Meredith continued, "We need to explain, but not explain away everything until we can no longer sense a pulse."

Fig. 6.1 The most recognized and lauded Pine Barrens blue hole, Berryland, NJ.

Inskeeps Blue Hole's eerie blue cast is now only apparent after ample summer rains reactivate springs. Berryland, near Folsom, NJ.

Fig. 6.2 A little known but notable Pine Barrens blue hole, Mount Misery, NJ.

This feature has virtually escaped modern perception, as have other Pinelands blue holes.

Fig. 6.3 Blue holes as potential biomes.

Unusual assemblages of cryptogams and algae have been noted at Mount Misery (Bien 1999) and at Inskeeps Blue Holes (personal communication, Walter Bien, June 2006).

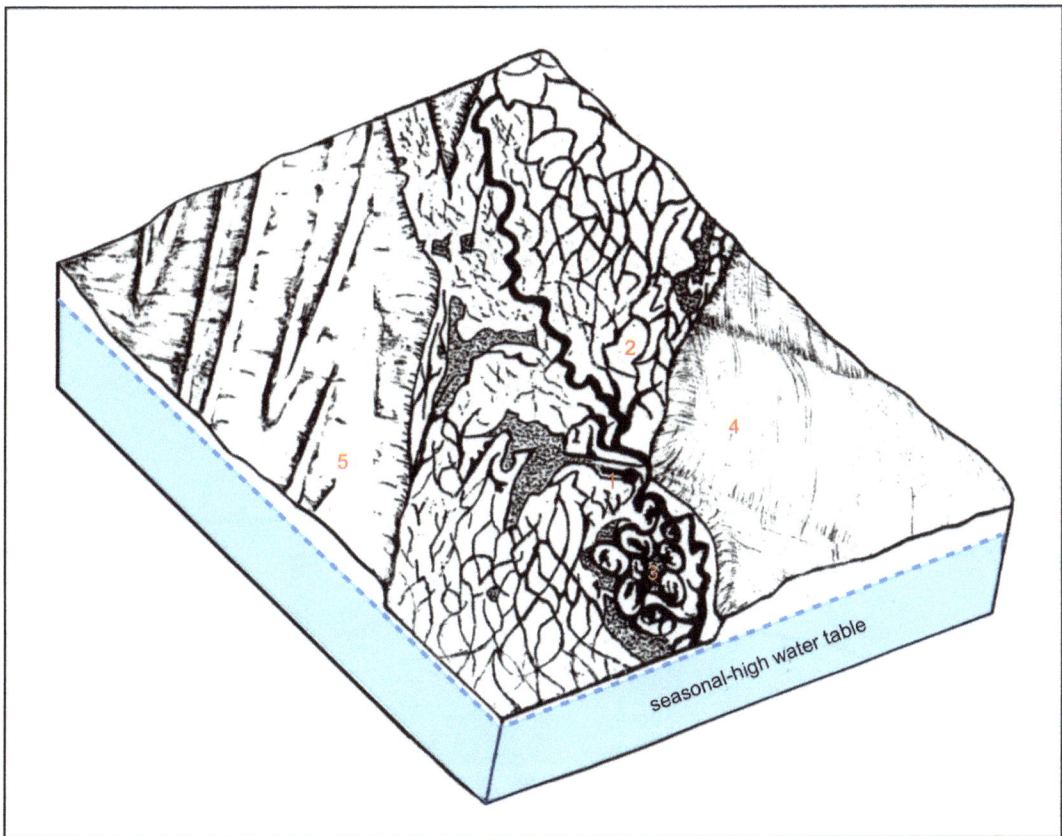

Fig. 6.4 Diagrammatic illustration of a blue hole setting, Berryland, NJ.

Inskeeps Blue Hole's geographic surroundings include: 1) a large circular spring; 2) a broad braided Pleistocene nival-flood paleochannel; 3) meanders downstream of the blue hole; 4) adjacent relict gravel bar providing the hydraulic head for a spring; and 5) an increase in sandy terrain below the spring where paleodunes and river meanders appear. Drawing by Pat Demitroff.

A UPWELLING WITHIN MODERN CHANNEL

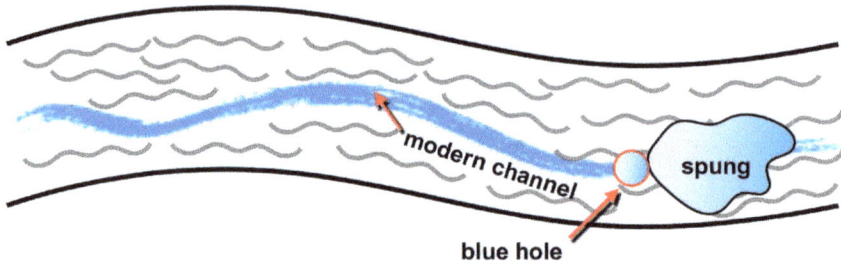

B UPWELLING WITHIN PALEOCHANNEL

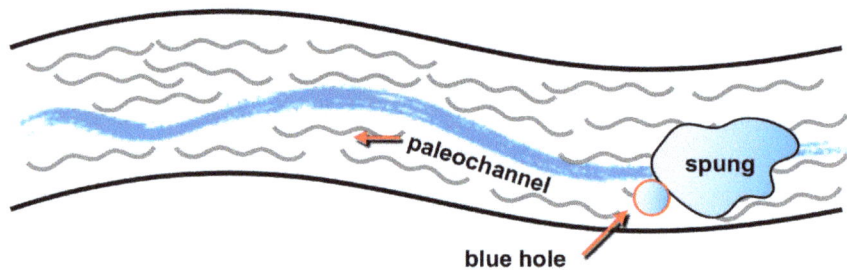

C UPWELLING WITHOUT PRIMARY LOCUS

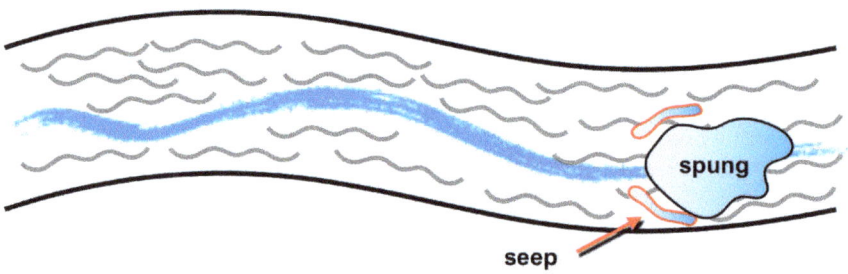

Fig. 6.5 List of 'blue-hole' spring forms observed in South Jersey associated with riverine spungs.

Some blue hole locations are associated with downstream riverine spungs. The springs are interpreted as ancient icing and aufeis sites.

(A) Upwellings with an obvious spring hole present within a modern stream channel.
(B) Upwellings with an obvious spring hole present upon a Pleistocene paleochannel.
(C) Upwellings without an obvious spring hole present but fed by seeps.

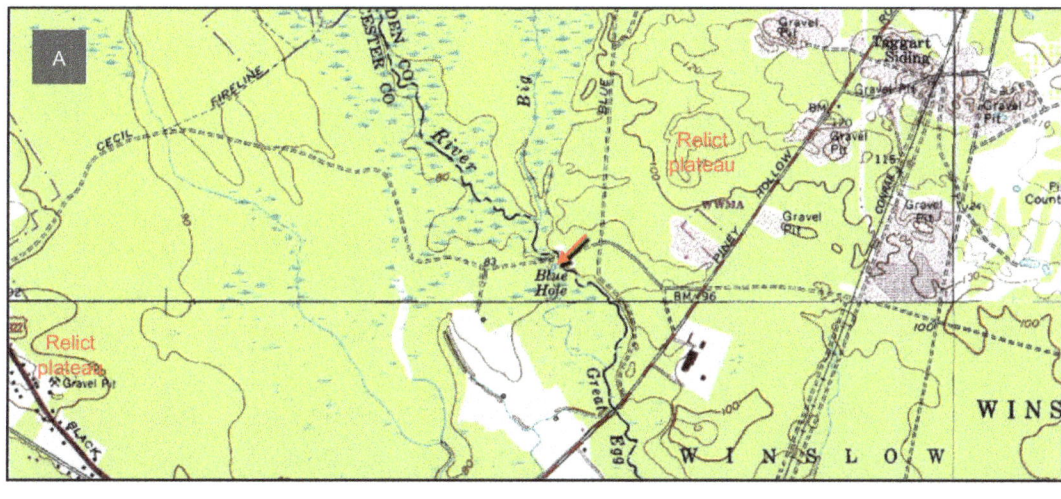

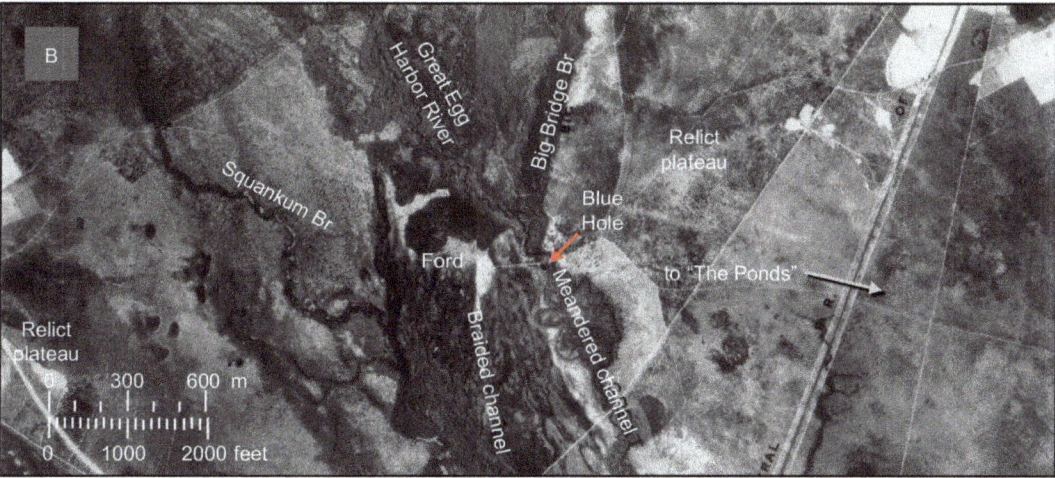

Fig. 6.6 Geographic setting of a paleochannel blue hole, Berryland, NJ.

Inskeeps Blue Hole is adjacent to the Great Egg Harbor River, which flows N–S.

(A) This blue hole's location (red arrow) is prominently labeled on the Williamstown 24k Topographic 7.5 Minute Quadrangle Map. Much of the groundwater-bearing relict plateau to the east of the blue hole has been mined away. Map imagery (USGS & NJOIT-OGIS, accessed through Map Info for USGS 24k Topo, Boyd Ostroff).

(B) The blue hole (red arrow) is clearly seen as a circular pond along with its ford where gravel bar remnants shoal the paleochannel as the Blue Anchor trail. East of the ford is the Shamong trail, which passes a series of spungs once known as The Ponds (Hider 1745, with "fresh meadows") where it intersects with the Long-a-Coming trail 4.8 km (3 miles) below Blue Anchor (Mattocks 1762, with "ponds"). Hider and later Mattocks owned Blue Anchor Tavern (Farr 2022), suggesting The Ponds and its meadows might be a way station for tavern operations. Similar ponds can be seen on c. 1931 aerial photomosaics at the Blue Anchor Tavern site, again the junction of ancient trails. Aerial imagery (NJOIT-OGIS c. 1931, accessed through 1930s Aerial Photography of New Jersey, Boyd Ostroff).

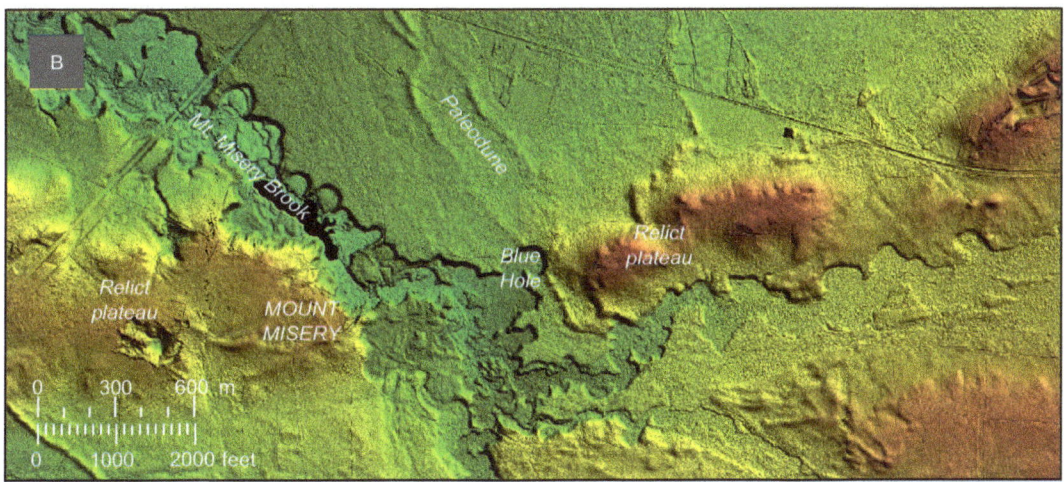

Fig. 6.7 Geographic setting of a paleochannel blue hole, Mount Misery, NJ.

Mount Misery Brook, adjacent to the Mount Misery Blue Hole, flows E-W.

(A) This blue hole's location (red arrow) is denoted as a distinct blue dot on the Browns Mills 24k Topographic 7.5 Minute Quadrangle Map. Much of the groundwater-bearing relict plateau to the east of the blue hole remains intact. Mount Misery, NJ. Map imagery (USGS & NJOIT-OGIS, accessed through Map Info for USGS 24k Topo, Boyd Ostroff).
(B) Lidar imagery indicates that this blue hole resides on a Pleistocene meander terrace that through sapping has incised itself into an older alluvial braidplain. Lidar imagery (USGS & NJDEP, accessed as Lidar in the Pines 2022, Ostroff 2021).

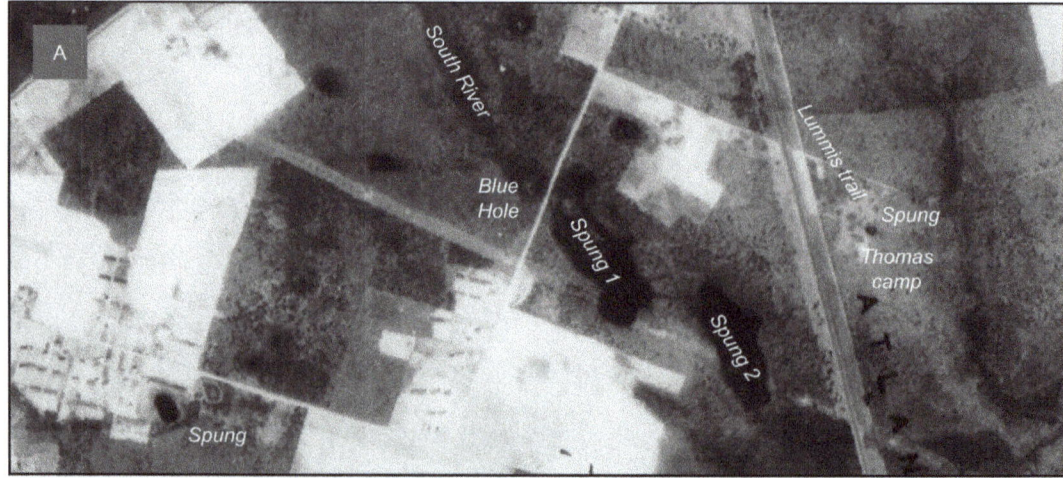
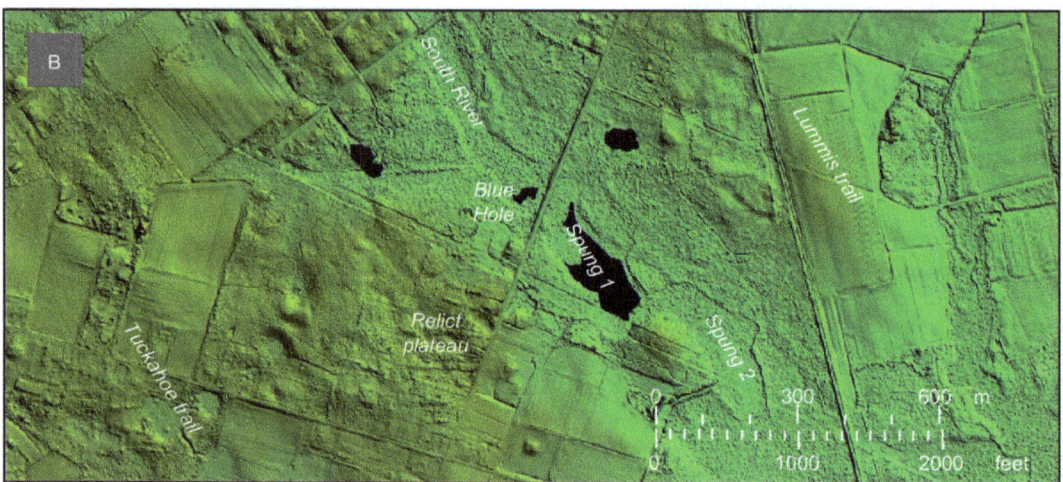

Fig. 6.8 Geographic setting of a modern-channel blue hole, Richland, NJ.

Danger Hole, atypical to most upwellings, is associated with two downstream riverine spungs.

(A) The Skating Ponds—designated as Spung 1 and Spung 2—are immediately downstream of the Danger Hole, a blue hole. Nearby, Thomas' c. 1870 charcoal and hoop pole camp was centered around a small round spung along the Lummis trail on its way to Lummis Pond (French & Demitroff 2001: 340, Fig. 2B). This author lives adjacent to (NW of) the camp. Richland, NJ. Aerial imagery (NJOIT-OGIS c. 1931, accessed through 1930's Aerial Photography of New Jersey, Boyd Ostroff).
(B). The width of the Skating Ponds is templated to its paleochannel banks. The Tuckahoe and Lummis trails bound from spung to spung in linkage, as do most early trails. Lidar imagery (USGS & NJDEP, accessed as Lidar in the Pines 2022, Ostroff 2021).

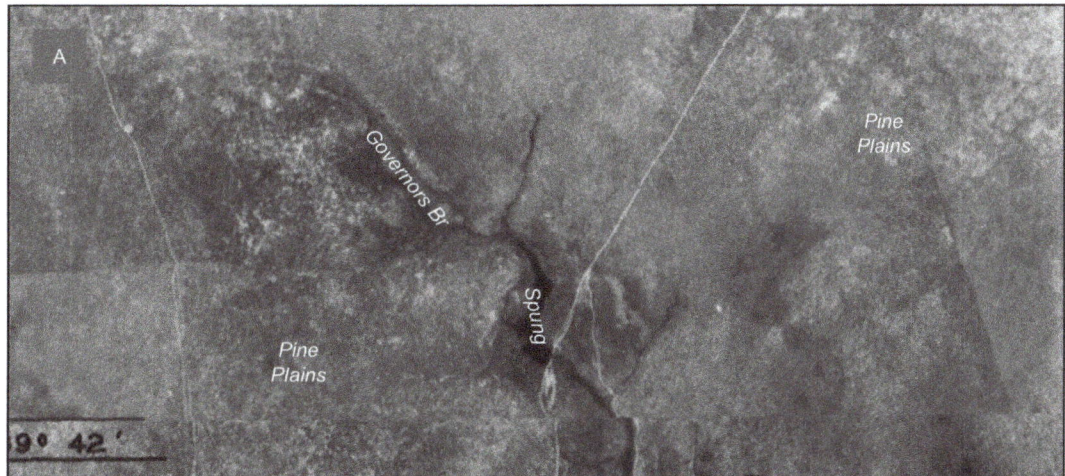

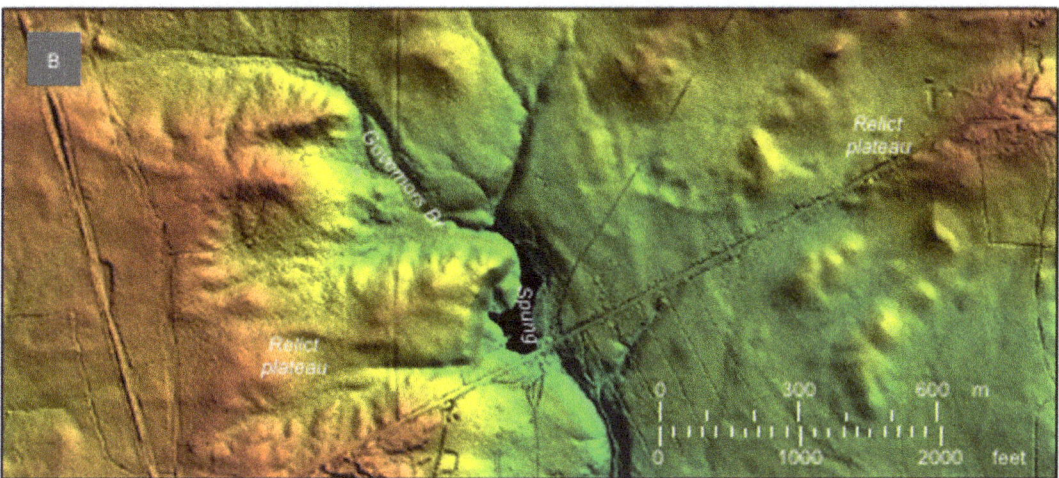

Fig. 6.9 Geographic setting of a hidden-seep blue hole, Warren Grove, NJ.

Watering Place Pond is an example of a named natural perennial riverine pond within a modern channel lacking an obvious blue hole. Harshberger (1916: 143–144) and Watts (1979: 448–449) studied Watering Place Pond.

(A) This pond was an important waystop deep in the pygmy pines of the East Plains. (NJOIT-OGIS c. 1931, accessed through 1930s Aerial Photography of New Jersey, Boyd Ostroff).
(B) Governors Branch has sapped back through seeps to its knickpoint at the base of its relict plateau spoon-shaped head. It is possible that this spung, too, would become intermittent today if were not for the modern damming of Governors Branch along Bombing Range Road. (USGS & NJDEP, accessed as Lidar in the Pines 2022, Ostroff 2021).

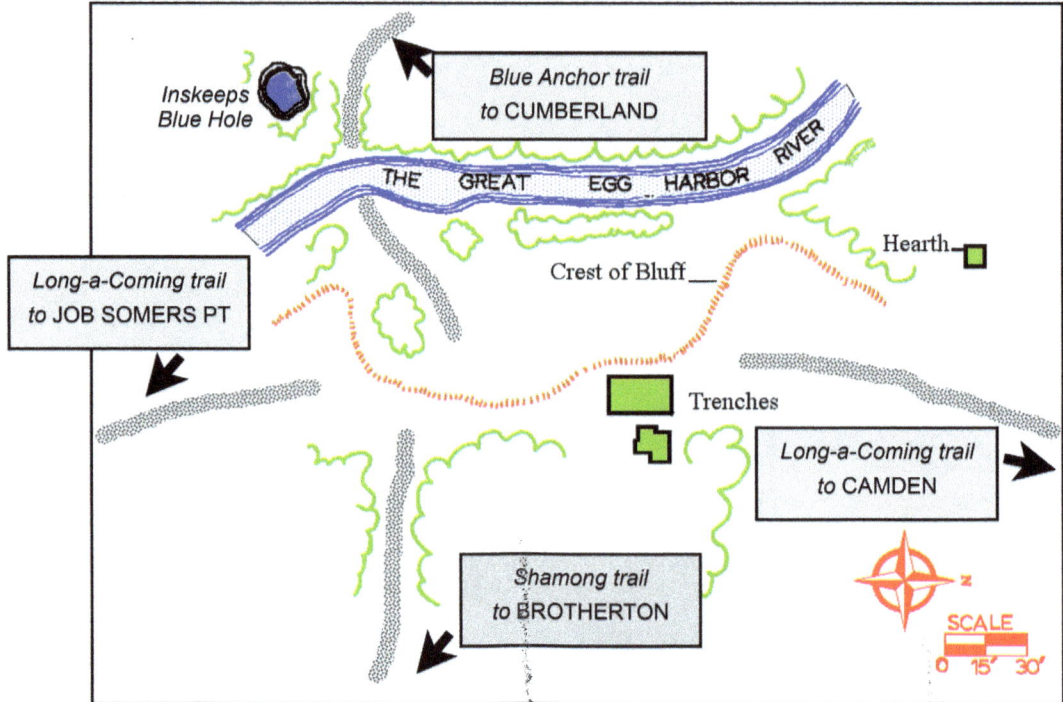

Fig. 6.10 Blue hole topography and precontact cultural use.

This diagram illustrates the relationship between a riverine bluff, a ford, and an ancient upwelling at the Inskeeps Blue Hole site. Precontact cultural artifacts are found in association (Cresson et al. 2006; adapted from Mounier 1972: 14, Fig. 1). Berryland, Folsom, NJ.

Fig. 6.11 Inskeeps' landscape is surprisingly complex in geomorphology.

(A) A striking change in paleohydrology occurs at the Blue Hole. Above it, the Great Egg Harbor River has a broad, high-energy braided paleochannel typical of ephemeral nival (snowmelt) flooding over frozen ground. There is a marked change to that ephemeral nival flood patter as it is quickly replaced by a meandering channel hydrologic regime. Aerial imagery (NJOIT-OGIS c. 1931, accessed through 1930s Aerial Photography of New Jersey, Boyd Ostroff).

(B) When nival flooding occurred, the ground was frozen, which included the bottoms of the river channels. Tabular river ice, called aufeis, was produced by a more perennial flow from the active spring or "icing" at the blue hole. The aufeis remained fast-frozen to the channel bottom when nival flooding occurred, ice damming the Great Egg Harbor River below the Blue Hole. In response, floodwater carved an auxiliary channel (in blue-dash line) around the ice dam and its adjacent relict plateau in capture of the Penny Pot Branch. Auxiliary channels are commonly formed in response to aufeis blockage (Åkerman 1982; Froehlich & Slupik 1982) and thermal-erosional tunneling (Alekseyev 2015). Relicts of such channels are commonplace along the larger waterways of the Pine Barrens (e.g., Fig. 5.11B, at Lummis Run). Meanders, too, can be generated in response to aufeis transgression (Alekseyev 2915: 196).

Note the true location of Three Pond Branch, the water course that empties into the Hospitality Branch below the junction of Three Pond Branch and Money Run. The lower course was once identified as Gold Run (McCarty 1757; Clement 1867: 25). This may be near the site where in May 1748 John Ladd Jr. visited a reputed gold mine near the Indian Orchard, where he dined—a tavern? (Stewart 1918: 72). He purchased a tract of land at the Indian Orchard, which was located at the head of Manaway Branch at The Lake (Ladd 1748), a circular spung now disappeared. This parcel is about 10 km (6 miles) from Gold Run. This is also near the site of the Huff and Jones blue hole reported earlier.

Silver mines, too, appear in local lore along waterways (Anonymous 1908a, 1908b; McMahon 1980: 106). Their long-abandoned shafts are often water-filled. This author suggested this lore too may have roots with ancient periglacial features like icings and thermokarst-like suffusion pits. The latter are common structures along Arctic riverbanks where frozen riverbanks have been thermally eroded (niched) from beneath by floodwaters (Melnikov & Spesivtsev 2000: 321; Woo 2012: 75, Fig. 3.2). This author has seen what appear to be natural pits along the high banks of the Great Egg Harbor River that are similar to suffusion pits. Lidar imagery (USGS & NJDEP, accessed as Lidar in the Pines 2022, Ostroff 2021).

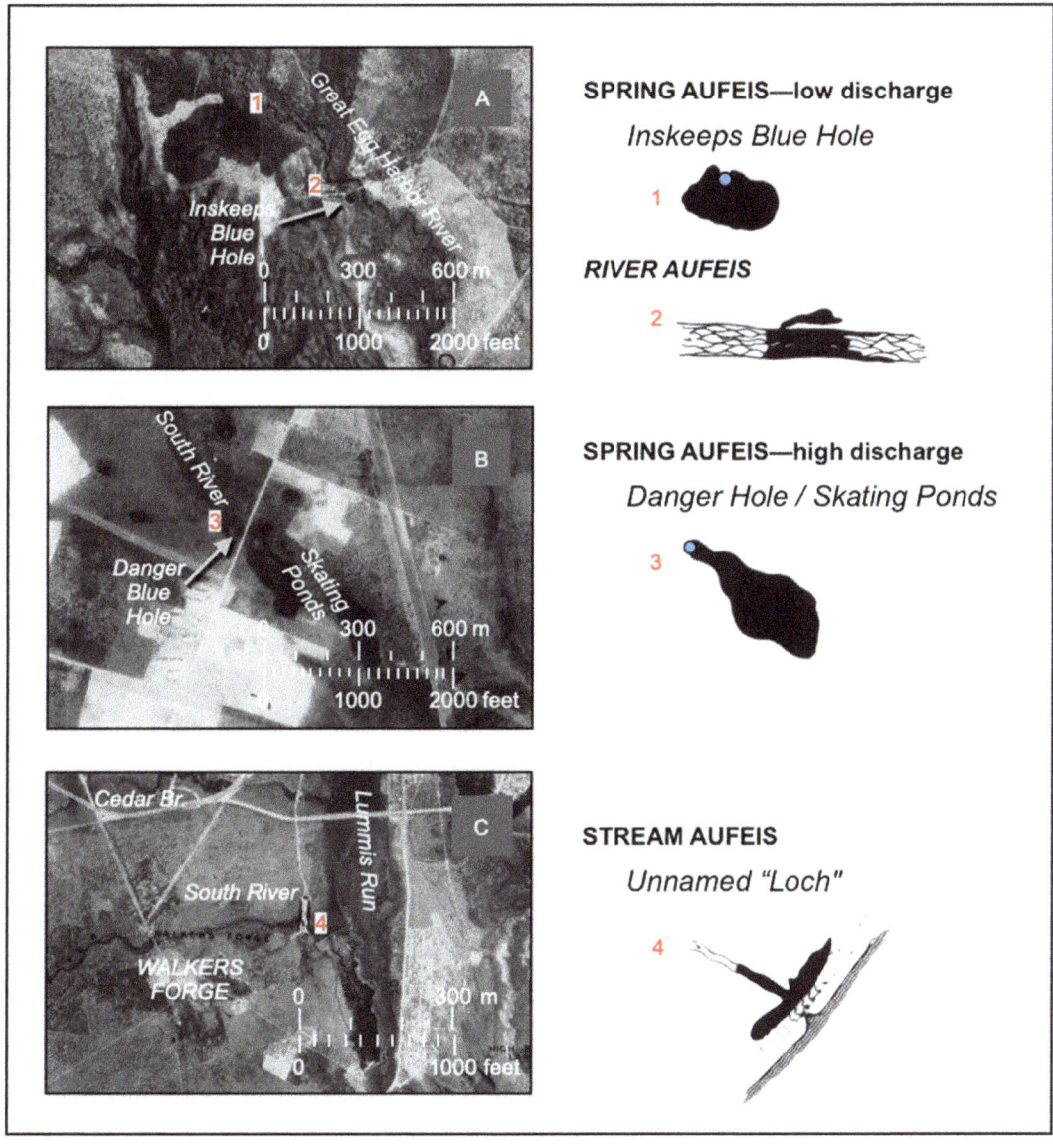

Fig. 6.12 Comparing riverine spung forms to suggested modern Arctic analogs.

Early aerial photos showing depressions (riverine spungs) in the Pine Barrens and their apparent similarity to aufeis in Spitsbergen (Åkerman 1982: 194, Fig. 4). Aerial imagery (NJOIT-OGIS c. 1931, accessed through 1930s Aerial Photography of New Jersey, Boyd Ostroff).

(A) Spring aufeis (1) of low discharge rate and river aufeis (2) with a spring of high discharge rate are compared with Inskeeps Blue Hole wetlands. Berryland, near Folsom, NJ.
(B) Spring aufeis (3) of high discharge rate is compared with the Danger Hole at the

Skating Ponds, Richland, NJ.
(C) Stream aufeis (4) is compared with bogs at the junction of the South River and an abandoned paleochannel (Lummis Run) of the Great Egg Harbor River at Walkers Forge, Mays Landing, NJ. Clement's (1867: 25) mapping suggest the pond here was loch-like.

Fig. 6.13 Source-bordering Pleistocene dunes found association with Pine Barrens lowlands, Millville, NJ.

Pine Barrens dune fields develop in accord with fluvial activity occurring close to their dune-building sand source (Wolfe et al. 2023). Spungs, cripples, blue holes, and savannah are wetlands. As such, they are found in lowland environments where dunes and coversands, too, form.

This is a view of a typical Pleistocene dune field near the Union House, some of which is visible as "sugar sand" alongside NJ Route 55 by the Millville/Vineland municipal border. Newell (2000: Sheet 2, [Qe]) identifies the location as dunal. Extensive patches of eolian sediments were chronicled in Newell's field notes, although due to stiff review he included only the most incontrovertible eolian deposits in his published geologic map (personal communications, Wayne Newell, March 22, 2006). Tedrow (1963: 7–8) noted "good field and laboratory evidence" of widespread eolian activity in the Pines. Throughout southern New Jersey, patches of loose frosted "sugar sand" can "set" (entrap) vehicles fast in their tracks (McPhee 1968: 60).

Fig. 6.14 Pleistocene paleodunes are abundant along ancient river terraces.

Here, at the Lochs-of-the-Swamp, the sand source is abundant—so much so that the ancient dune crests reach heights of up to 4 m (13 feet). This crest elevation is near the height limit for that of Pinelands paleodunes. The location is an active dune study site (Wolfe et al. 2023). Long narrow spungs or "lochs" formed between the long dune arms. Lidar imagery (USGS & NJDEP, accessed as Lidar in the Pines 2022, Ostroff 2021).

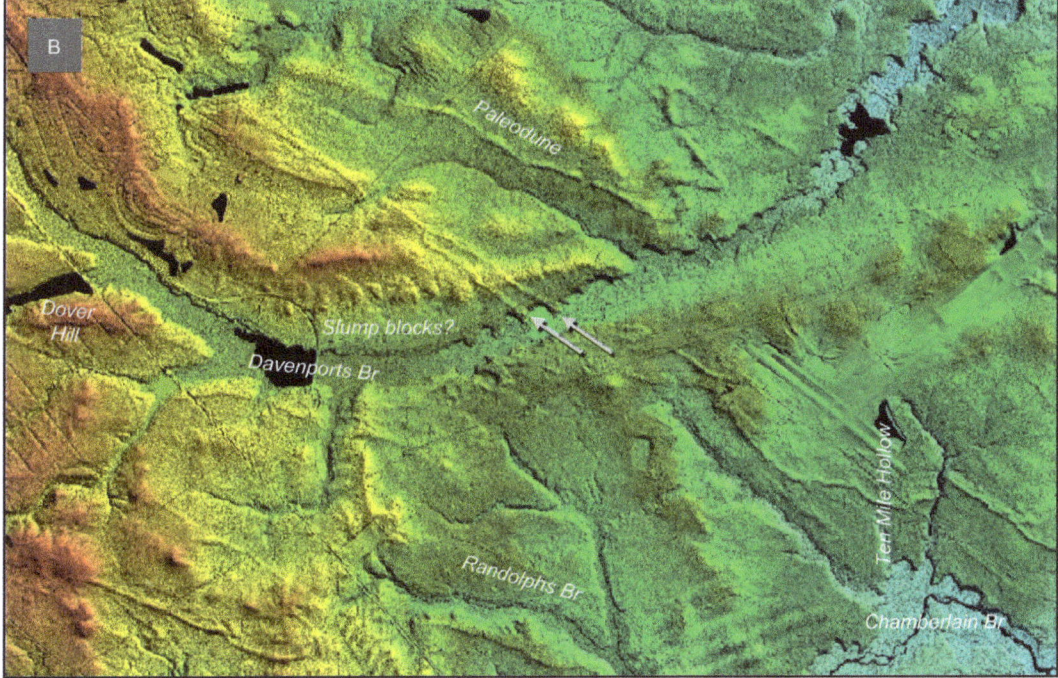

Fig. 6.15 A rare wind streak feature reaches across the central Pine Barrens.

Its apparent trace lies between Whiting and Waretown, NJ. Although common on Mars (Sagan et al. 1972; Cohen-Zada et al. 2016), it is a terrestrially rare feature ordinarily constrained to the driest deserts. Zanner (1999) reported its possible extant in ice-marginal Minnesota.

(A) Early aerial photos, taken when deforestation was maximum, shows unusual lineations of dark and light sand that extend approximately 14.5 km (9 miles) from northwest to southeast across the central Pine Barrens. Preliminary work suggests these streaks are composed of dark and light magnetized sands of eolian origin. There is no elevational difference between the dark streaks and the light streaks. However, the darker bands bear higher magnetic charges (Demitroff 2010; French & Demitroff 2012). Note how the Davenports Branch changes its channel-form downstream W–E: 1) braided channel, no wind streaks—thin coversand activity; 2) megameanders, wind streaks appear—thickening coversand; and 3) micromeanders, wind streaks disappear—thicker coversand. (NJOIT-OGIS c. 1931, accessed through 1930s Aerial Photography of New Jersey, Boyd Ostroff).

(B) Laser altimetry reveals what appear be dislocated slump blocks on the Davenports north bank, possibly due to thermal erosional niching during nival flooding similar to the block slump reported by Newell and DeJong (2011: 270) in Maryland. Note how two parabolic dune arms align their form to meander cusps (at arrows). (USGS & NJDEP, accessed as Lidar in the Pines 2022, Ostroff 2021).

07
Savannahs of the Pine Barrens

Etymology of Savannahs

In modern usage, savannahs are grasslands located in tropical and semitropical regions with distinct wet and dry seasons (OED 2012). For example, in Africa, they are large, open plains of tall grassland, with a scattering of drought-resistant trees. The term "savannah," as used here, is from its proper derivation (OED 2012). Caribbean in origin (OED 1989, given by Oviedo 1535 as a Carib word), the word was used by the Indigenous population and Spanish explorers to describe sparsely treed, open, alternately wet and dry meadows and plains of tropical America. The Spaniards soon exploited them for cattle production. By the eighteenth century, in a North American sense, the term referred to wetland meadows along inland streams and within Carolina Bays (Meredith 1731a, 1731b). In the modern (nineteenth century and later) manner of speech, savanna was increasingly used to describe open plains with long grass. In Pine Barrens narrow-sense use this author recommends the traditional *savannah* spelling to describe wetland or riverside savannah, to distinguish it from the upland application of the spelling *savanna* as the two ecosystems are very different landforms and biomes.

Brown (1879: 3, 10), Whitehead (1875: 129–130) and Smith (1765) give early accounts of cattle raising on southern New Jersey savannah (Fig. 7.1). Mountford (2002: 112–113) adds sheep and horse to island livestock (also see Smith 2020). Coastal savannahs were sometimes called "salt prairies" and had numerous freshwater springs that were continually "boiling and bubbling" throughout where stock could water (Rose et al. 1878: 21).

Early on timber and cattle were important Pinelands commodities (Purvis 1920). According to early accounts, cattle were raised in the bogs and meadows of the Pine Barrens (Figs. 7.2–7.4), but the surrounding soils were too poor for cultivation (Brown 1879: 15; Blackman 1880: 406). Although it is well documented that salt marsh savannah made fine pasturage (Thomas 1940: 300–304; Wilson 1953: 137–140; Smith 2020), little is known of the use of the Pinelands' savannahs. During the colonial period, cattle were driven from wooded pasturage along the Great Egg Harbor River to feed upon shore savannah every winter (Wrangel 1969: 11–12). By the turn of the nineteenth century, local inhabitants only rarely used the inland savannahs for pasturage (Gifford 1900: 248).

In Buckhorn, an eighteenth-century Swedish naval store hamlet between the railroad-era villages of Richland and Milmay on Calf Branch (Taylor 1880a, Calf Pasture Branch),[1] the Tomlin family continued to pasture cattle on a seasonal basis into the 1890s. This practice ended with the advent of the Richland & Petersburg Railroad, which consolidated into South Jersey Railroad in 1893 (Cook 1966). By the family account, "Many cattle were killed, for they would get upon the railroad tracks" (Tomlin 1932: 33). This account would place grazing in and around a spung-cluster known as the Lee Ponds[2] on Green Branch[3] (Hackney 1795) (Fig. 4.5B).

---

1. This author grew up on a poultry farm at the Calf's head and played around the farm duck pond—a spung—that began Calf Branch, which has since dried up. This writer in-play building forts found much joy uncovering buried campfire remains, odd stone chips, and an Archaic "blood-point" chert spear point near this depression, once observing a ball-lightning event at the site. Early on it was considered a special place in our value system. On May 05, 1980, the author's brother hosted an impromptu music festival for a couple hundred fellow Richard Stockton College students. The mud pit was plowed as a de facto spung near but not in the head pond, so as not to in any way degrade the existing Calf Pasture Branch basin.

2. A geologic account of Lee Ponds is offered in Chapter 5, "Origins of Spungs."

3. There are numerous Green Branches in the area, no less than eight within about 15 km (10 miles) of Buckhorn. Most are small-order streams bordered with marshland. In frequency they are only succeeded in occurrence by generic ever-present Cedar Branches. It is possible that Green is not a person's namesake, but a land-use description. Green is regionally not a common early family name. By Green's frequent occurrence there should be Greens in association with the designated Branches. In contrast there are Ingersolls at Ingersoll Branches, Sharps at Sharp Branches, Seeleys at Seeley Branch . . . , but few if any Greens. It is therefore proposed that Green could relate to grazing ground. According to the *Oxford English Dictionary* (OED 2011) green—among other meanings—as chiefly British refers to "A piece of public or common grassy land situated in or near a town or village, from which it often takes its name; a village green," or generally "Grassy ground; a grassy spot (now chiefly *British regional*)" or chiefly in

Tomlins also raised cattle in the woods in Jefferson, Gloucester County, west of the Pine Barrens. Buckhorn resided in old Gloucester County until Atlantic County formed from it in 1837. In a letter to Tomlin (1932: addendum), a relative recounts, "The cows had bells on and pastured in the woods. Sometimes they got mired in the mud."

Ted Gordon, botanist, researched the term and interviewed numerous Pinelands residents to aid in its definition. Early on, *savannah* was used to denote meadows. With the advent of cranberry growing, it is now variously applied to lowlands with leatherleaf (*Chamaedaphne calyculata* [L]) and pitch pine (personal communication, Ted Gordon, December 21, 2006). It is possible that the latter application could be delineative of former meadows that have since become woody. Townsends Swamp (Fowler & Lummis 1880) (Fig. 8.9) and Atsion Meadows/Grassy Lake (Bisbee 1971: 258) were once considered savannah habitat (Harshberger 1916). In *Cranberry Culture* (White 1885: 32–35, 43–48), savannahs were interpreted as level, hummocky, sparsely wooded, and intermediate wetlands separating swamps from wetlands.

When describing wetlands, it is easy to quagmire in semantics. Waksman (1942: 16–56) considers scores of terms for New Jersey's wetlands, and defaults to bog, a catchall term in common use (e.g., cranberry bog). Fen was first used by McCormick (1955), but later abandoned in his reevaluations (1970: 53–58; 1979: 230–235) for the more traditional savanna(h), spong (and spung), and cripple wetland designations. Collins and Anderson (1994) argued that bogs are specifically peat-accumulating, nutrient-poor wetlands without water inflow or outflow. They redefined South Jersey's savannahs and spungs[4] as fens since they are wetlands *with* water inflow and outflow. Walz et al. (2006) and Palmer (2005)—following Collins and Anderson—abandoned savannah and used fen to describe long and narrow riverine wetlands. Smith (2012: 5) continued, equating savanna(h) with fen. This was an important distinction since

---

word modification "a piece of grassy ground used for a particular purpose" or "Green vegetables or other plants used to feed animals." In example, the Buckhorn/Lee Place Green Branch was exploited for grazing. Just above Savannah Branch on the Maurice River appears Green Branch and Swamp (Penn 1750) The two branches are of similar sandy-marshy geomorphologic constitution. Nearby and closer to The Lake was The Park (Chew 1815), a savannah-like marsh along Manaway Branch above Cow Horn Branch (Hartman 1978: Map 5). Green Branches may also be coppice sites for woody forage (Ingrouille 1995: 211, 252, 253).

4. Collins and Anderson (1994: 159–160) suggested that Pinelands' wet depressions (i.e., spungs) and savannahs are anthropogenic.

160 SOGGY GROUND

these wetlands, unlike bogs, have leaky ecosystems with complex inputs and outputs.

Bog and *fen* are peatland terms, applied to wetlands that contain a minimum of 40 cm (16 in) of partially decomposed organic matter (National Wetlands Working Group 1997: 1). McQueen (1990: 17) states, "A fen is a *Sphagnum*-dominated wetland . . ." and discusses fen dynamics in depth (1990: 29–31). The wetlands of the Pines share some characteristics with poor fens (Gignac et al. 2000: 1140, open-system groundwater, acidic conditions), but nearly all lack the thick *Sphagnum* deposits associated with peatlands. Gignac et al. (2000) concluded that New Jersey, because of climatic factors, is too far south to have *Sphagnum*-dominated peatlands (see footnote 24 on peat). Palmer (2005: 4) explains why he characterized savannah as fens, "These wetlands are *approximately considered* fens due to their status as peatlands with consistent groundwater input . . . pH, cation concentrations, and other nutrient inputs are generally *quite low* . . . placing these fens *at the poor fen end* of the minerotrophy gradient . . . somewhat similar to the fens in the Midwestern United States." Palmer cautiously uses *fen* qualifiers like "approximately considered" and "somewhat similar," as savannah reside on the fringes of what we think of as a fen.

Thick deposits of true peat are said to exist in some lower terrace savannah sites (Walz et al. 2006, to 240 cm), but that is not the norm for upper terrace savannah sites where thick peat is uncommon. That would mean some but not all parts of a savannah can be fen-like. Thus, it is suggested that the original geographic term *savannah* better suits this landform in its entirety as it includes fen-like and non-fen-like conditions.

Appellatives of Savannahs

Few specific appellatives containing "savannah" have been found in the Pine Barrens. A watercourse known as Savannah Branch in the fringe area of "Broad Neck"[5] (namesake of the Broad Pond spung, Figs. 4.6D, 4.22) has a grassy terrace where deer are abundant. It is the first of three savannah habitats encountered traveling eastward along the Cohansey trail (*see below* Early Use of Savannahs and Topography). Just east of Broad Neck on Manaway Branch is "The Park" (Chew 1815) below "The Lake" (Fig. 4.5B) by Blue Bell Tavern, suggesting that savannah habitat

---

5. See Chapter 8, "Spungs as Indicators of Hydrologic Change": Broad Pond.

had once occupied this place. Another example is Alquatka Meadows and Branch, for which Atco, New Jersey, is named (Farr 2002: 4–5). A place called Long Meadows once existed in the Penny Pot Branch watershed head near "The Ponds" (Bishop 1831) with "fresh meadows" (Hider 1745). Paul (1822) identifies a savannah pond near Pomona, New Jersey, used in a generic descriptive sense. Old Oxbow Fen, Big Doughnut, and Long Savanna are savannahs ("fens") investigated by Palmer (2005), but no deep provenance has been found for these names. In a study of savannah Smith (2012) termed these subdivisions based on ecological habitat succession: 1) wet savanna; 2) graminoid savanna, and 3) shrub savanna.

Morphology of Savannahs

The forester Gifford (1900: 245) commented that the names for various wetlands of the Pinelands were "exceedingly elastic in meaning." He applied "savannah" to herbaceous wetlands, with few or no trees. He attributed the presence of the savannah to a dense layer of compacted material or "hardpan" found just beneath the ground surface. This hardpan was considered analogous to the "ortstein"[6] of northern Europe, or the "alios"[7] of the French Landes (Gifford 1900: 248).

Pitch pines, which grew on scattered knolls, would often topple because of the poor rooting atop this shallow hardpan (fragipan?).[8] Gifford felt this land was nearly worthless. Its potential for forestry was low because, even if drained, the indurated layer would inhibit tree growth. Because the investment for preparing and draining a savannah for forest

---

6. Ortstein is a hard-cemented soil horizon that restricts root penetration (Schaetzl & Anderson 2005: 443). It is related to bog iron ore (Raseneisenstein). The former is attributed to podzolization and the later to gleying processes (Kaczorek et al. 2004). In the Pine Barrens, bog iron ore is attributed to iron-oxidizing bacteria (Braddock-Rogers 1930: 1495–96; Means et al. 1981).

7. Alios is an indurated podzolic horizon rich in organic matter (Jambert et al. 1997: 106).

8. Spung bottoms are high-energy sites of intense Pleistocene cryogenic quartz weathering (Demitroff et al. 2007; French et al. 2009b), and less so with cripple and savannah bottoms. Their floors become dense like fragipan, but lack the characteristic desiccation cracks associated with typical brittle fragipan (cf. Demitroff 2016: 126, Fig. 2, floor of modified wedges and sediment-filled pots). Fragipan genesis process through physical and chemical weathering remains enigmatic (Bockheim & Hartemick 2013). Although savannah habitat is often associated with a fragipan-like bottom, such a layer may not be present. The full role of savannah-bottom substrate warrants further investigation.

production would never be recouped, he concluded that they were "destined to remain unchanged for many years to come" (Gifford 1900: 248).

Harshberger (1916) questioned application of "savannah" to the Pinelands meadows. He stated, "True savannas, such as exists in the southern United States, in subtropic and tropic countries . . . [in the strictest sense] do not exist in New Jersey" but later reasoned that they could be called savannahs by applying a physiognomic test. Like a true savannah, New Jersey's analog is "a flat, grassy area or plain, treeless or dotted over with clumps of trees or individual trees scattered over the surface." He found it difficult to clearly separate a swamp from a savannah. Harshberger limited these grassy wetlands to one or two flat terraces (1916: 147, Fig. 122) found along valley watercourses.

Apparently, savannahs once covered a much larger region of the Pine Barrens (McCormick 1970; also see Plate I, inside back cover pocket, Harshberger 1916; Palmer 2005; Smith 2012). The range of savannahs shown in Harshberger (1916: Fig. 122) appears to correspond to the data present on an earlier topographic map series (Cook & Vermeule 1890: Sheets 15 & 16). Smith (2012: 54, 63) noted that his savanna[h] study area declined by 71.3% in coverage between 1940 and 2002, with a "wet savanna" loss of 89.4%. Their disappearance may be related to lack of natural or human disturbance (Dowhan et al. 1997; Smith 2012: 63-64, also river damming & extirpation of beaver) and/or the diminution of regional groundwater, discussed in Chapter 8 of this work.

Savannahs occupy broad paleochannels associated with ancient drainageways (Demitroff 2003), often forming long, narrow meadows (50 m to 1+ km long, ~50 m wide/164 feet to 3280+ feet long, ~164 feet m wide) along Pinelands river corridors (Walz et al. 2006; Palmer 2005: 2-3; Smith 2012, terms them "riverside savanna"). Savannah habitat can be separated into three paleochannel typologies: a) braided paleochannels with an adjacent well-defined modern channelway; b) braided paleochannels without an obvious present-age channelway; and c) meandered paleochannels (Figs. 7.5–7.10).

Early Use of Savannahs and Topography

Ancient aboriginal trails often pass along savannah habitat. The Cohansey trail (Purvis 1917: 27-29; Hampton 1917: 49-50) crosses the Savannah Branch in Pittsgrove Township, the Townsends Swamp savannah (Harshberger 1916: Plate 1) in Buena Vista Township (Fig 8.9), and later meets the Long-a-Coming trail with three large narrow ponds at the Lochs-of-the-Swamp (Figs. 4.9C, 4.10). This trail juncture is the location of the c. 1706 Steelman Plantation,[9] and at the southeastern end of the Weymouth savannah. The Steelman family apparently had a large number of cattle listed in their early inventories (New Jersey Secretary of State 1734, 1736).[10] Habitat at the Lochs-of-the-Swamp provided abundant venison for hunting (Wrangel 1969: 13).

Concurrent with work on enclosed basins, the study of prehistoric cultural dune associations developed within the last few decades, with initial investigations or periglacial landscapes initiating during the 1970s. Much of this pioneering research is centered upon the flanks of Inner Coastal Plain cuesta mounts,[11] and eastward onto the Outer Coastal Plain. Large-scale Late Pleistocene (and Holocene?) source-bordering dune fields are common but under-recognized by the scientific community in this state. However, recent work has been undertaken to reveal Coastal Plain dune dynamics south of the Pine Barrens (Markewich et al. 2015; Swezey 2020). Eolian landscapes are under active study by this author and colleagues (e.g., Demitroff et al. 2019).

Source-bordering dunes are usually local in sediment source, related to the deflation of proximal waterways in the case of parabolic dunes and wetland depressions in the case of lunettes that rim some spungs. During Pleistocene cold periods, wetland landforms like paleochannels and closed basins were often ephemeral features that allowed wind entrainment of sand grains during their dry phases. Prehistoric use of Pinelands dunes is poorly understood. Even less is known about the hairpin parabolic dunes bordering savannah habitats (sedgy/grassy tree-sparse wetlands occupying paleochannels). Some dune features have "special place" use such as burial sites and ritual landscapes; others exhibit the more mundane patterns of ephemeral foraging encampments (Fig. 7.11) (Cresson et al.

---

9. This property was conveyed by Thomas Budd to James Steelman (~1695) along the Great Egg Harbor River and is mentioned by Clement (1880: 414; *also* Hess 1910: 4), and like other early pioneers neglected to officially record their titles. Pastor Wrangel visited the Steelman plantation in October 1764 (Wrangel 1969).

10. The "Inventory of James Steelman" from 182H: Eggharbour January 4th 1734/5 lists some provocative possessions among his possessions, "one neigro man & three Indian squaws" and "68 Head of Cattle" along with "five fatt [pregnant] cattell." The author believes that the Lochs-of-the-Swamp savannah at the plantation was grazing land for the earliest Pinelands inhabitants despite the dominance of generally unpalatable sedge.

11. Cuesta is used as per Wolfe (1977: 277) and Mounier (2008) and cuesta mounts as per Wolfe (1977: 285) and Mounier (2003: 157; *also see* Demitroff 2016: 125).

2006; also see Abbott 1918). Aboriginal cultures are known to locate their villages at dune margins. Dry or crusted sands act as a moisture-holding mulch that collects rainfall, providing a source for springs along dune edges (Greeley & Iversen 1985: 158; Tsoar & Møller 1986: 91). Paleodunes are also present in tideland savannahs (Figs. 7.12–7.13), where they provide rises to pass upon.

As with spungs, cranberry growers also modified savannahs to grow their "red gold."[12] The flat fragipan-like floor beneath the hummocky fluvial and eolian sands (White 1885: 33, Fig. 7) was an excellent template to follow. The planar hard surface is remarkably level due to a periglacial inheritance, facilitating the filling and draining of bogs. The densified bottom-layer also provided an aquiclude or aquitard that conserved water during bog-fill.

Savannahs were also opportune locations for iron ore mining. Bog iron (Russell & Krug 1980; Means et al. 1981) was the raw material for South Jersey furnaces, an important Pine Barrens industry of the eighteenth and nineteenth centuries (Boyer 1931; Pierce 1957). Many furnaces (e.g., Weymouth, Atsion, Hampton, Martha, Speedwell) were located near savannahs (Fig. 7.12) (Braddock-Rodger 1930: 1496, meadows). Another name for bog ore is "meadow ore" (Blackman 1880: 416). In savannah habitat, there are minimal numbers of tree roots to remove for ore exploitation. Roman and Good (1983: 44) suggest savannah-type areas are the result (at least in part) of bog ore mining. Meredith (1731a) described naturally occurring savannah along inland rivers of North Carolina's "Pine Barrens," indicating New Jersey's savannah habitat too probably also predates intensive settlement by Europeans. This assumption is further supported by Southgate (2000: 8) who concluded through pollen that the type vegetation at the savannah study site had persisted for millennia.

Of Savannahs and Plains

Some confusion exists about current usage of *savannah* in the Pinelands. Sparsely wooded uplands are occasionally also referred to as *savanna(h)s*. The name suggests open, park-like dry woods, perhaps

---

12. Calling cranberries "red gold" (common in the Pine Barrens) is attributed to Charles Dexter McFarlin, a Cape Cod, Massachusetts, native who went West to mine yellow gold, came home to grow cranberries, then returned to Oregon to pursue mining for red gold instead (Hall 1942).

akin to the plains of Africa (*see* "Etymology of Savannahs" in this chapter). In New Jersey, the term has historically been used with reference to a grassy wetland with a scattering of pines or whitecedar. Open tracts on uplands, with low pines and stunted growth, are referred to as "plains,"[13] not savanna(h)s (Figs. 7.15-7.16) (Gibson et al. 1999). Now pretty much limited to the Pinelands core area (Good et al. 1979), extensive tracts of the scrublands once were scattered throughout the Pine Barrens (Engle et al. 1921: Plate II; McCormick & Buell 1968: 24) (Figs. 7.16-7.17).

The name "Belleplain" designates the area where a large scrub plain once existed (Gordon 1828; 1850; Cook 1857, "the brushland") in Upper Township. The Risley Plain in Weymouth Township became part of the Estelle Colony (Risley 1896). Around Newtonville, the now extinct Heath Hen was abundant in the "barrens of Gloucester," a plain-like region straddling Buena Vista Township, Folsom Borough, Winslow Township,[14] and the Town of Hammonton (I. 1832: 15-16). This variant of the Prairie Chicken of the Middle West was usually associated with the habitat of the East and West Plains of Ocean and Burlington Counties. Although never officially reported from Cape May County, Stone (1937: 320-322) suspected that it had been present at an earlier time. Certainly, a plains habitat capable of supporting Heath Hen had been present there in the nineteenth century.

Terrain identified as supporting Pinelands "plains" habitat is composed of broad stretches of level or moderately undulating ground that is elevated and underlain by "hardpan" (Pinchot 1899: 125-130; Harshberger 1916: 166, 223; Good et al. 1979: 288-289). In a model of landscape evolution for southern New Jersey (Newell et al. 2000, Fig. 4; French et al. 2007: 55, Fig. 2), Pine Plains would be limited to the highest elements of the block diagram (Fig. 7.18). A close relation exists between surficial materials, topographic elements, and how the surface has been affected by mass wastage.

The densified brittle nature of the B/C soil horizon is attributed to a long period of leaching and subaerial oxidation (Owens et al. 1983). In uplands in and around the Pine Plains, Trela (1984: 558-562) suggested

---

13. Plain, like spung, is in use by seventeenth-century settlers to designate landforms along the sandy flats of the Nashua River in northern Massachusetts. Green (1912, p. 62—Brown Loaf Plain & High Plain, both flat, sandy, and pine-covered; p. 63—Pine Plain; p. 64—Tobacco Pipe Plain). Other Pine Plains are the Rome Sand Plains, Chase Lake Sand Plains, Saratoga Sand Plains, these in New York (Lookingbill et al. 2012).

14. Hollinshead (1779) cites a place called Eryens Barrons in what is now Winslow Township that was in about the area where the Gloucester Plains was thought to once be.

that B/C soil-horizon microfabrics (i.e., oriented argillans, papules) were indicative of intense desiccation/cryogenic deformation and these highly oriented clays cemented the hardpan. Reinterpreted as a (periglacial) "fragipan,"[15] this hardpan-like layer played a large part in preserving sand- and ground-wedge casts from deformation upon the thaw of permafrost (French et al. 2005; Demitroff 2016: 132, Fig. 6). Towards cast preservation it is probable that the fragipan armored the upper ground surfaces from landscape modification when permafrost degraded at the end of the cold periods. During thaw the hard fragipan resisted ice-melt deformation during permafrost dewatering and increased saturation from precipitation with climate amelioration—especially when the newly formed groundwater was perched by a cryogenic aquiclude or aquitard.

Many (mostly Europeans) suggest that a fragipan is at least in part related to a periglacial climate or permafrost (FitzPatrick 1956; Lozet & Herbillon 1971; van Vliet-Lanoë & Langohr 1981; French et al. 2009a), although there are other hypotheses to explain its genesis (Bockheim 2014: 164). Fragipan-like (fragic) brittle layers, also found beneath spung bottoms, and dense fragipan-like (fragic) layers beneath cripples, paleochannels, and bottoms of deformed wedges and sediment-filled pots of the Pinelands (Demitroff 2016). Either they are too recent in origin to have experienced a long period of leaching and subaerial oxidation associated with a weathering fragipan model, or their wetter locality somehow inhibited brittleness and desiccation cracking.

Origin of Savannahs

Little (1946: 37, 39) suggested that dry-season wildfires formed and sustained the wetland meadows. If a forest fire consumed the woody tops of plants, but left the organic mat relatively unchanged, woody plants would regrow rapidly. However, if the organic layer was consumed down to the mineral soil, regrowth of woody vegetation would be delayed until a *Sphagnum* mat of sufficient thickness accumulated to form a new seedbed (Little 1979: 308–310). Whittaker (1979: 320) agreed that fires during the dry season, combined with soil factors, were the likely causes

---

15. Basically, hardpan when wet remains dense but will bend. Fragipan when wet does not bend but instead slakes along cleavage planes. For more on the definition and identification of fragipan and fragic see Bockheim (2014: 153; *also* Witty and Knox (1989: 2–4)

of savannah-like conditions. Southgate (2000: 8) concludes that hydrology and fire history, not human factors, are key to savannah presence.

Farrell et al. (1985) examined an area of complex braided debris fans at the Atsion Ore Bogs, a swampy area that once supported a more widespread savannah community (Cook & Vermeule 1890, Sheet 15; Harshberger 1916, Plate 1). From the sedimentological evidence, Farrell stated that the braided valley represented high-discharge fluvial events, perhaps due to the debouchment of a sediment-choked Delaware River (Marsh 1985; Farrell et al. 1985: A-10). Later, these fans were reinterpreted as braided paleochannels resulting from the thaw of permafrost (Newell & Wyckoff 1992; Newell et al. 2000), and reflect high seasonal (i.e., snowmelt-derived) flow over seasonally or perennially frozen ground, combined with high sediment transport (French & Demitroff 2003: 123; 2012). Due to the absence of exotic materials in the sediments (Newell & Wyckoff 1992), a Pleistocene debouchment of the Delaware River was discounted. The Atsion fluvial braid-plain terrace is often underlain by a sparse pebble lag that contains ventifacts (Demitroff 2016, incl. pavement einkantes). During fluvial activity the terrace had to be frozen rock-hard during the brief snowmelt/ice-melt flood events, which allowed the formation of broad-braided channels in sandy terrain. Hairpin parabolic dunes bound over about a meter (three feet) of eolian and fluvio-eolian sands, which is mantled on silty gravelly colluvium. The basement is composed of Cohansey sands.

Using the Cook and Vermeule (1890, Sheets 15 and 16) and Harshberger (1916, Plate I), it appears that savannah habitat occurred in three types of paleochannels in the Pine Barrens (Fig. 7.14). These ancient channels are somewhat akin to fluvial systems in tundra barrens or polar semi-deserts found in sandy areas on islands of the western Canadian Arctic today (French & Demitroff 2003: 123; 2012). There the broad-braided channels are the result of brief snowmelt/ice-melt flood events occurring over frozen ground during a short summer.

In the European literature there are numerous references to the hydrological implications of paleochannel pattern changes resulting from climate change (Starkel et al. 1991; Frenzel 1995; Vandenberghe 1995; 2008; & Woo 2002; Murton & Belshaw 2011; Gębica et al. 2015). The study of river valley paleohydrology is called paleopotamology (Rotnicka & Rotnicki 1988). When correlated with sedimentological and chronometric data, a convincing story can be told of Late Wisconsinan to early Holocene fluvial dynamics. European braided river environments are often associated with ephemeral, high-energy flow over frozen ground during the coldest phases of the Weichselian (=Wisconsinan) (Mycielska-Dowgiallo 1977; van Huissteden et al. 1988; Mol 1997; et al. 2000; Kasse 1998; Huisink 2000; Zielinski & Gozdzik 2001; Kiss et al., 2015). Very large river meanders in Europe are frequently ascribed to climate amelioration during the Late Pleistocene/Early Holocene transition (Gonera 1986; Vandenberghe et al. 1987; & Sidorchuk 2020; Szumanski & Starkel 1990; Starkel 1995; Sidorchuk et al. 2000; Kasse et al. 2003; 2010; Borisova et al. 2006).

A transition from a braided to meandering river channel at the end of the Wisconsinan was also reported in the eastern United States (Leigh 2006; 2008; et al. 2004; Suther et al. 2018). This change in channel pattern was considered unrelated to glacial or periglacial conditions, but instead was a response to changes in vegetation density. Given the nature and widespread occurrence of periglacial features in the Pines, it is likely that some fossil channel forms are also preserved there from coeval frozen-ground conditions. In permeable sands and gravels, it is difficult to ascribe Pine Barrens wide flat-bottomed waterways to conditions other than frozen ground (Kozarski 1995: 89), particularly under temperate interglacial hydrologic regimes such as the current system present in today's Holocene where streams are highly underfit[16] to their accommodating paleochannel terraces.

Ancient dunes are often in association with savannah habitat (Figs. 7.3, 7.12). Little is known of Pinelands dune chronology. Initial reconnaissance near Indian Branch[17] (Demitroff 2010) indicates that these deposits are

---

16. According to "Neuendofr et al. (2005: 696), a underfit stream is "A misfit stream that appears to be too small to have eroded the valley to which it flows; a stream whose volume is greatly reduced . . . It is a common result of drainage changes made by capture, by glaciers, or by climatic variations."

17. The patterned ground contained within the paleochannel of Indian Branch is unusually well defined at this location. Dunes dammed the channel in the past, their nearby "noses" still visible stacked upon each other as seen on lidar imagery (Demitroff et al. 2019), and this windblown channel fill is dated to 28.16 ± 2.80 ka BP (Demitroff 2010, Demitroff et al. 2012). A modern stream channel incised the ancient impoundment along both the north and south paleochannel walls. When sand-choked, it is surmised that water would have impounded at Indian Branch. When the water body drained, the sediments were ice-rich, which facilitated frost cracking. In the Arctic, lakes artificially drained by Mackay (1999) quickly developed thermal contraction wedges in their newly exposed bottoms (*also see* Walters 1983). In arid polar regions, patterned ground is often limited to depressions, which are rich in ice (van Vliet-Lanoë 2005: 292, Fig. 11.60 c.), and alluvial plains in sandy-gravely terrain (van Vliet-Lanoë 2006: 86). During the dry, sparsely vegetated Pleistocene, ice-wedge polygons were limited to waterlogged sites (van Vliet-Lanoë 2006: 87). The potential for preservation of Pleistocene microfloral and microfaunal material at this site and

primarily Late Pleistocene in origin because its sands have undergone obvious pedogenesis since their deposition, which is indicative of a long period of soil weathering. Six optical dates from the work of Wolfe et al. (2023) suggest that hairpin parabolic and transverse dunes stabilized from 23–17.5 ka. Typically, Pinelands dune features are dismissed as "fire shadows" (Gibson et al. 1999: 60) on aerial photos. Works on inland eolian deposits in Europe (Kozarski 1991; Szczypek & Waga 1996; Schirmer 1999; Dulias & Pelka-Gosciniak 2000) indicate that dunes are episodic, and often reactivated during the Late Pleistocene and Holocene. Similar ages have been ascribed to dunes in the eastern United States (Denny & Owens 1979; Markewich & Markewich 1994; et al. 2015; Ivester et al. 2001; & Leigh 2003; French & Demitroff 2012; Swezey et al. 2013; 2016; 2020), but overall studies are limited in comparison to the extensive European works on Pleistocene dunes.

Many areas of sugar sand have a windblown origin, especially when proximal to western and southern banks of watercourses—source-bordering dunes.[18] These are interpreted as proximal or riverine dune sediments, predominantly entrained by the same katabatic winds that deflated the enclosed wetlands (i.e., spungs) of the Pines (French & Demitroff 2001), creating the raised rims (lunettes[19]) found on the southeastern (downwind) margins of some spungs.

Wolfe et al. (2023) demonstrates through quart-grain morphoscopy and granulometry that their "skinny" sand-starved dune systems were subject to very intense in situ frost weathering under periglacial conditions. Dune sediments were the result of saltation and suspension transport. The New Jersey Pine Barrens may have been a particularly windy dusty place, particularly during the Mid-Wisconsinan.

---

other nearby dune-blocked streams may be high. Stone (1911: 799) noted this is a "wild spot" for plant collecting.

18. Frosted windblown sand grains collected from collected from periglacial macro-structures (French et al. 2007; 2009b) and from eolian channel fill (Demitroff 2016; Wolfe et al. 2023, dunes) is highly reported to be rounded by wind transport and display upturned surface fracture indicative of roundness and frosting due to bombardment during eolian events as per Woronko & Bujak (2018; et al. 2015). Also, strong bleaching is reported in the luminescence dating reported in previous two reports, which again support an invocation of eolian processes in association with these frosted grains.

19. Lunettes (Livingstone & Warren 1996: 101–102) are defined as wind-created crescentic mounds in arid and semiarid areas that are emplaced on the downwind side of playa-lake basins, forming in association with lowering waters in a fluctuating water table. Unlike a true dune (Neuendorf et al. 2005: 196), a lunette is not capable of movement.

There are modern accounts of dune-building in southern New Jersey (Scientific American Correspondence 1896: 183—East and West Plains; Anonymous 1893—Belleplain; Dillingham 1911: 106—Hammonton; Abbott 1918: 50—Trenton), indicating: 1) an abundance of loose surface sand; 2) that material is dry; and 3) there is an absence of a productive vegetation cover (Castel 1991: 117).

Savannahs: Summary

Palmer (2005) suggested that just a 10 cm (4 in) difference in savannah hummock elevation above groundwater can have a profound effect on vegetation type. Since these wetlands contain many rare or endangered plant species, he considered their conservation a highest priority. Smith (2012: 62) further argued that these Pinelands wetlands are unique ecological communities endangered by anthropogenic drivers like fire suppression and beaver extirpation. This author further suggests savannah habitat is extremely sensitive to changes in water table baseline, making this ecosystem a valuable barometer of groundwater change.

Complex small-scale environmental heterogeneity is considered an important attribute of savannah habitat (Palmer 2005). Much of this microtopogarphy is attributed to Late Pleistocene and perhaps early Holocene processes, a time when the broad cold-climate paleochannels were abandoned for temperate modern channels and dunes were anchored by vegetation as the climate ameliorated. A review of the Pinelands paleohydrology is a prerequisite for a temporal delineation of past and present savannah habitat. Additional examination of wetland meadow geomorphology will aid habitat conservation and restoration efforts. Interrelations between proximal dunes[20] and wetland features remain poorly understood and under investigation by this author and colleagues (Wolfe et al. 2023).

---

20. Warwick in Stewart (1932: 91) notes an Indigenous trail preference for the "warm or southern side of streams." Pleistocene sand dunes occur on the southern and eastern (downwind) banks of Pinelands waterways. Sandy substrate "dry trails" provided good soil drainage, warmed quickly, were more sparsely vegetated, white-frosted sugar-sand grains aided night travel, and the elevated deposition infilled low elevations providing easier passage over wet spots. Hunters relate that game follow dune crests because they can better see prey from the higher position, the scent of lurking prey below updrafts, and the loose dry sands leave less scent behind for prey to track.

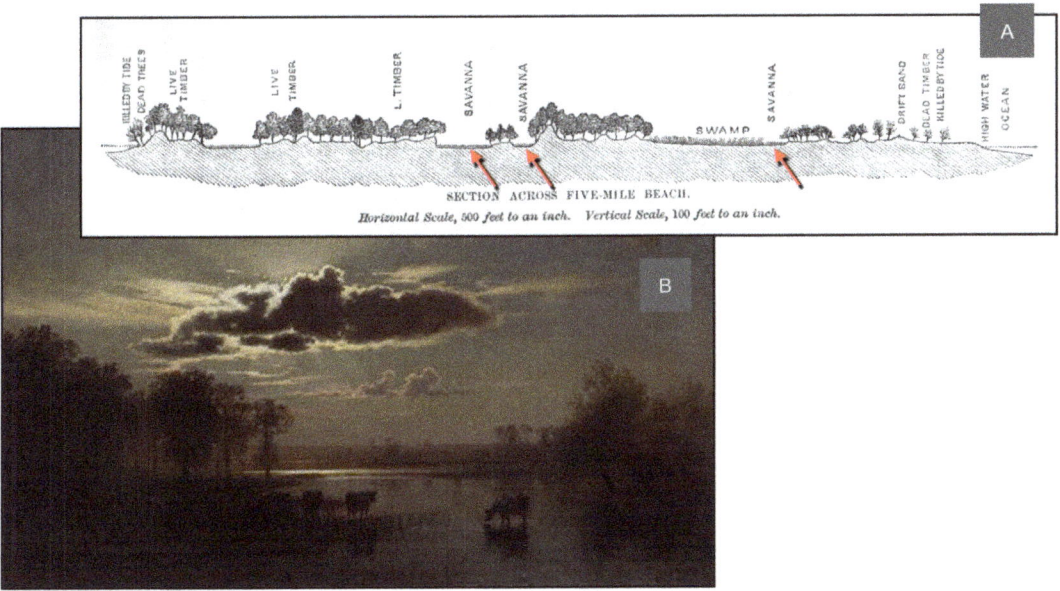

Fig. 7.1 Coastal savannah and cow meadow.

(A) Pine Barrens-like barrier-island savannahs. Interdunal meadows found on barrier islands along the Jersey shore were called savannah (red arrows) and utilized for cattle pasture by the earliest settlers in the seventeenth century. The diagram (Cook 1857: 52, Section 1) is an old cross section of Five-Mile Beach, known today as Wildwood. Note that the early Pine Barrens vegetational boundary bordered tideland along the coast (Fig. 1.1), hence an extension of Pine Barrens environment onto barrier islands.

(B) "Sunset on a Florida Marshland with Cattle Grazing," painting by Herman Herzog, 1898.

Fig. 7.2 Typical Pine Barrens savannah near Weymouth, NJ.

It is a vestige of once-widespread savannah between the Lochs-of-the-Swamp and the Weymouth Furnace ore beds. The meadow is dominated by Walter's sedge (*Carex striata* [Michaux]).

Fig. 7.3 Savannah/paleodune associations and interactions are commonplace.

An ancient pathway associated with the Long-a-Coming trail complex follows a dune crest through meadowland proximal to the Great Egg Harbor River and the savannah shown in Fig. 7.2. Long, narrow loch-like spungs near Weymouth are interpreted as interdunal deflation hollows. These "lochs" resemble canoe-shaped, flat-bottomed deflation basins of the Kobuk Valley in Alaska (Dijkmans & Koster 1990). In southern New Jersey, savannahs were often in association with wetland "badlands." Weymouth, NJ.

Fig. 7.4 Typical Pine Barrens savannah near Atsion, NJ.

This is a bog-iron savannah along the west bank of the upper Batsto River at Lower Forge. Hummocks in the foreground are coastal sedge (Carex exilis [Dewey]), with twig rush (*Cladium mariscoides* [Muhlenberg]) in the background. Photo by Ted Gordon.

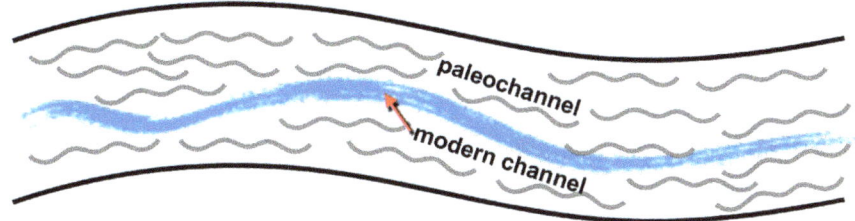

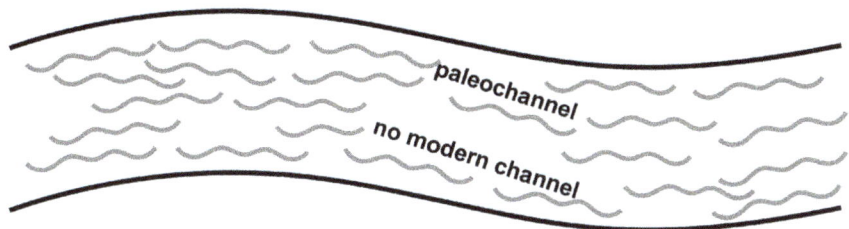

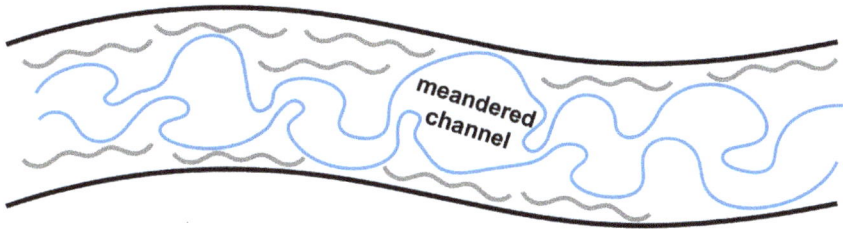

Fig. 7.5 List of savannah forms observed in the Pine Barrens.

(A) Some savannahs reside on a Pleistocene-aged broad braided river terrace or terraces aside a well-defined incised modern Holocene river system.
(B) Some savannahs reside on a Pleistocene-aged alluvial fan terrace or terraces without a well-defined incised modern Holocene river system.
(C) On occasion, savannahs reside on a Pleistocene-aged (Pleistocene–Holocene?) meander terrace or terraces.

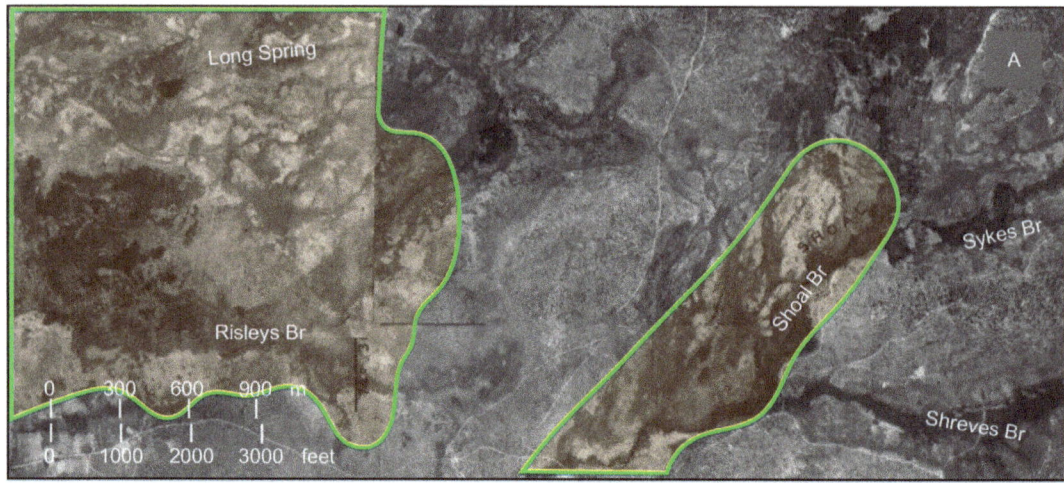

Fig. 7.6 Geographic setting of Types-A & -B savannah forms, Chatsworth, NJ.

Savannah boundaries are drawn from Harshberger (1916, Plate 1) as seen in Fig. 7.14E.

- (A) Type-A savannah form is outlined in green and highlighted along Shoal Branch (*right*). Type-B savannah form is outlined in green and highlighted bordering Risleys Branch and Long Spring (*left*). Aerial imagery (NJOIT-OGIS c. 1931, accessed through 1930s Aerial Photography of New Jersey, Boyd Ostroff).
- (B) Topographic map vegetation icons confirm the presence of savannah habitat with sparse pine and whitecedar. Map imagery (Cook & Vermeule 1890, Rutgers University, accessed as Cook's Map of the Pines, Boyd Ostroff 2019).

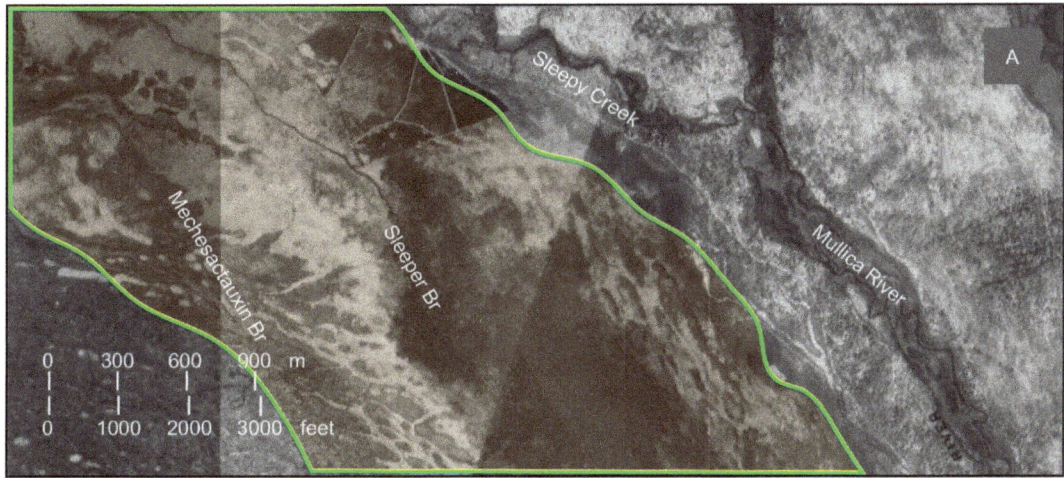
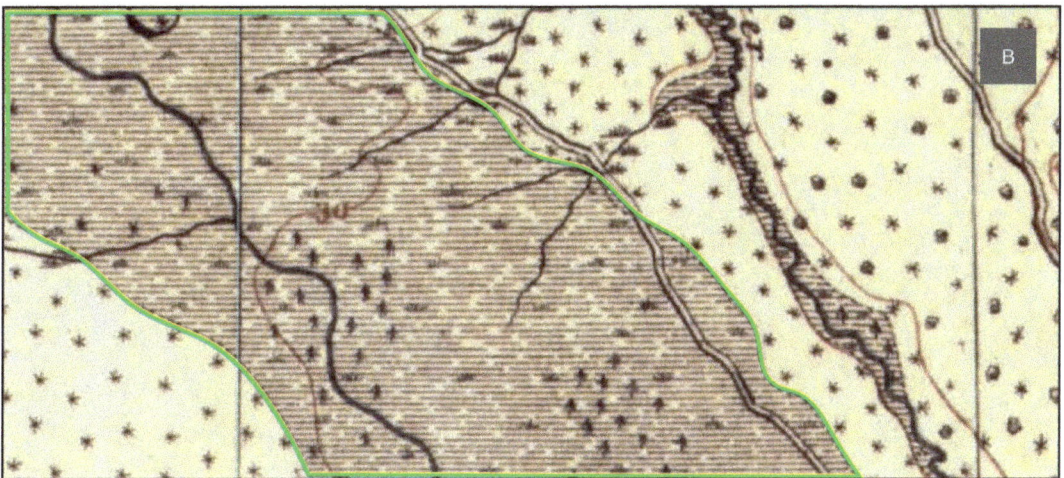

Fig. 7.7 Geographic setting of Type-B savannah form by Rockwood, NJ.

The site is north of Hammonton. Savannah boundaries are drawn from Harshberger (1916, Plate 1) as seen in Fig. 7.14A,B.

(A) Type-B savannah form is outlined in green and highlighted. The savannah resides on an alluvial fan. These are not river terraces associated with Mechesactauxin or Sleeper Branches. Aerial imagery (NJOIT-OGIS c. 1931, accessed through 1930s Aerial Photography of New Jersey, Boyd Ostroff).
(B) Type-B savannah form is again outlined in green and highlighted. Topographic map vegetation icons confirm the presence of savannah habitat with sparse pine and whitecedar inferred from their density. Map imagery (Cook & Vermeule 1890, Rutgers University, accessed as Cook's Map of the Pines, Boyd Ostroff 2019).

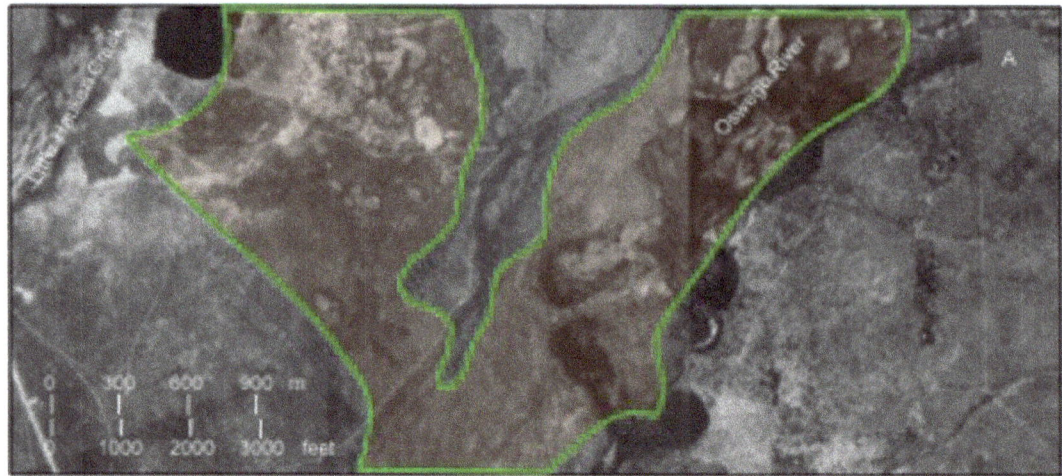
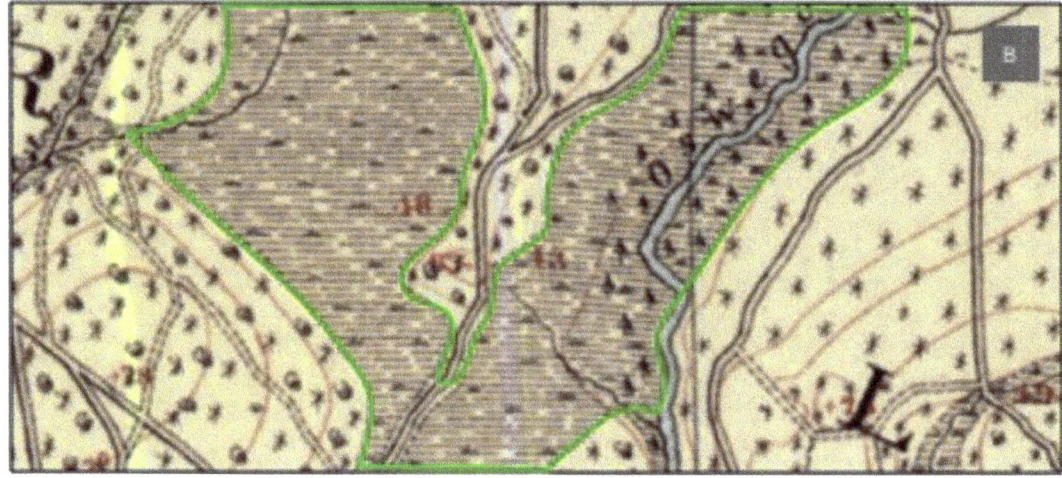

Fig. 7.8 Geographic settings of Types-A, B, & C savannah form Chatsworth, NJ.

The site is E of Jenkins and S of Chatsworth. Savannah boundaries are drawn from Harshberger (1916, Plate 1) as seen in Fig. 7.14D.
(A) Types-A, -B, & -C savannah forms are outlined in green and highlighted. Type-A savannah (center, west of Oswego River) resides on a wide braided terrace of the Oswego River. Type-B savannah (*left*) resides on an alluvial fan without a dominant incised modern waterway. Type-C savannah (*right*) resides on a meander terrace of the Oswego River that has incised into the older braided terrace that hosts Type-A savannah. Aerial imagery (NJOIT-OGIS c. 1931, accessed through 1930s Aerial Photography of New Jersey, Boyd Ostroff).
(B) Types-A, -B, & -C savannah forms are again outlined in green and highlighted. Topographic map vegetation icons confirm the presence of savannah habitat with whitecedar (esp. on the meander terrace) inferred from their density. Map imagery (Cook & Vermeule 1890, Rutgers University, accessed as Cook's Map of the Pines, Boyd Ostroff 2019).

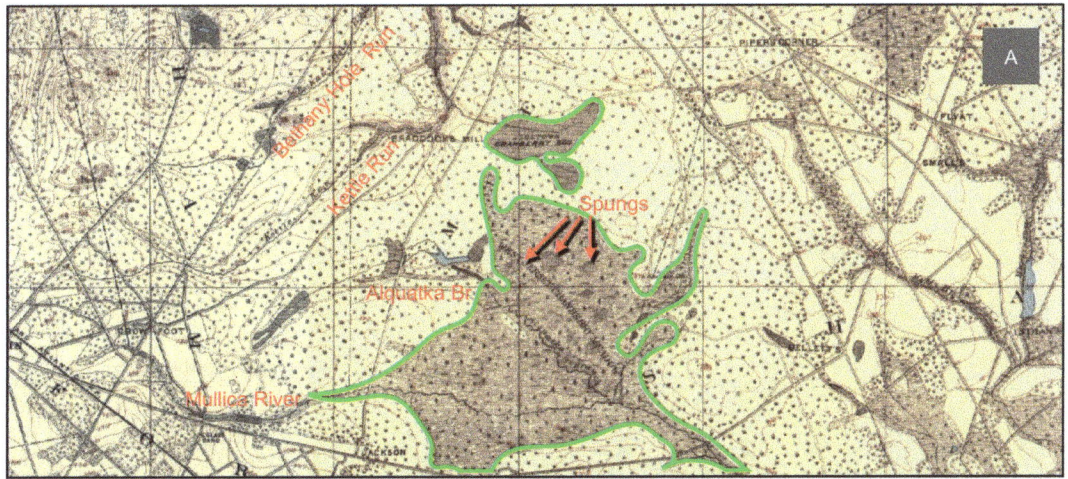
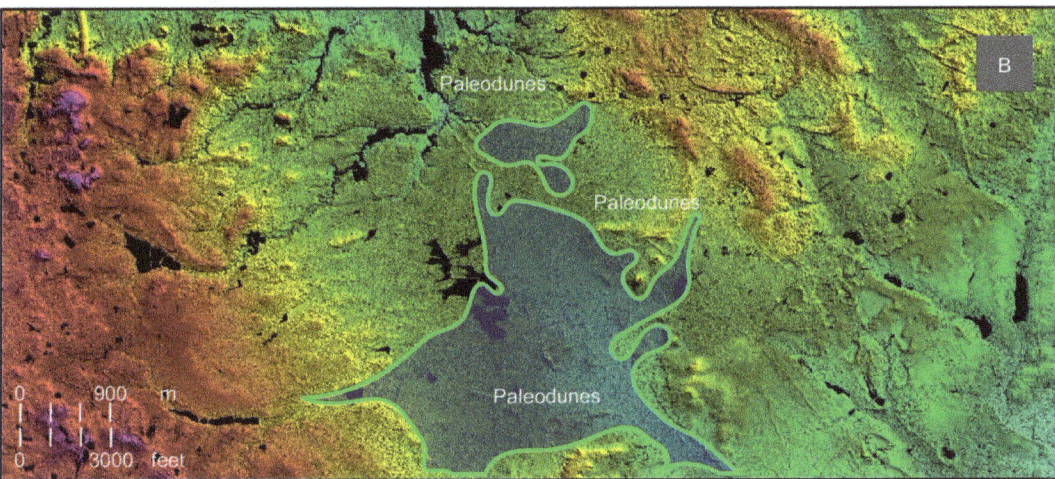

Fig. 7.9 Geographic setting of Type-B savannah form, Atco, NJ.

The site at Alquatka Meadows between Atco and Medford, NJ. Savannah boundaries are drawn from Harshberger (1916, Plate 1).

(A) Type-B savannah form is outlined in green and highlighted. The savannah resides on an alluvial fan that hosted a number of spungs. Topographic map vegetation icons confirm the presence of savannah habitat with sparse whitecedar inferred from their density, with a few pines in the west. Map imagery (Cook & Vermeule 1890, Rutgers University, accessed as Cook's Map of the Pines, Boyd Ostroff 2019).
(B) Type-B savannah form is again outlined in green and highlighted. Part of this wetland was used to establish the West Jersey Cranberry Meadow. Note the presence of paleodues, although it is likely many more were destroyed for cranberry production. This is the general location of archeological work at dunes, spungs, and savannah by Mounier and Cresson described in Fig. 7.11. Lidar imagery (USGS & NJDEP, accessed as Lidar in the Pines 2022, Ostroff 2021).

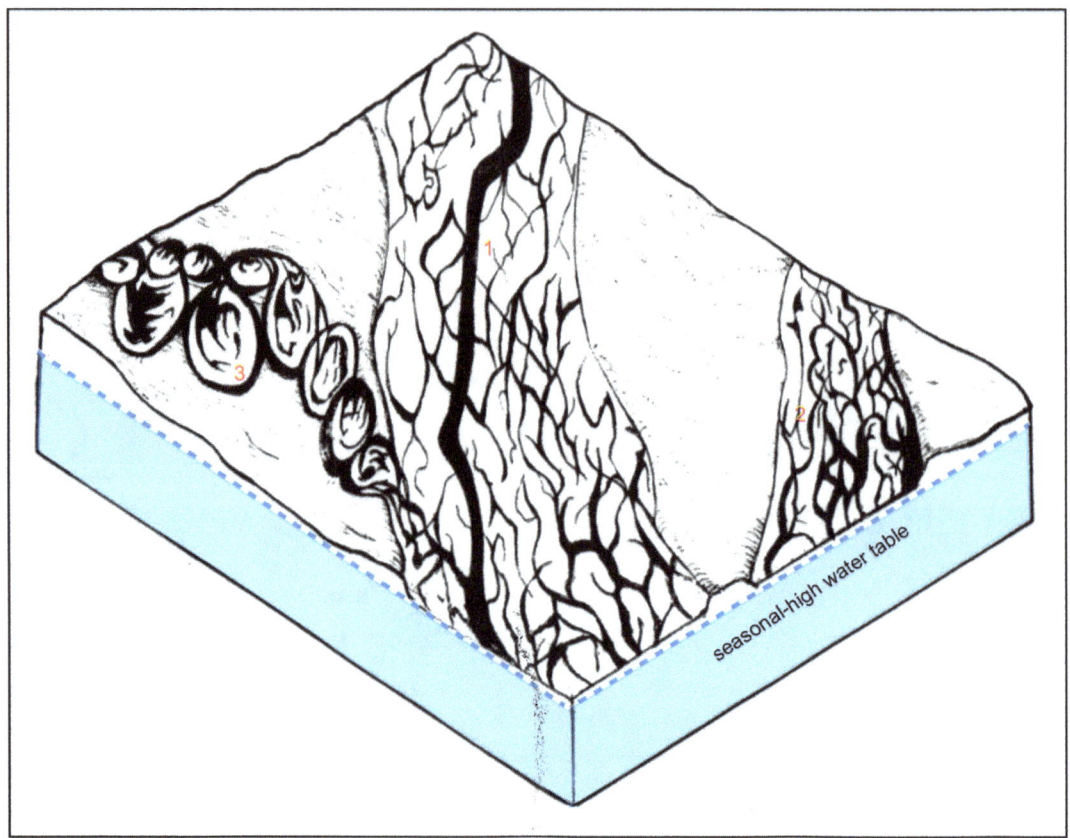

Fig. 7.10 Diagrammatic illustration of savannah settings.

1) Type-A savannah form—broad braided river terrace or terraces; 2) Type-B savannah form—alluvial fan terrace; and 3) Type-C form—meander terrace or terraces. Harshberger (1916: 146–150) records a two-stage terrace at the Wading River creating both a wet and dry savannah facies, the upper being Type-A and the lower Type-C. The latter's sandy, cuspate channels are younger, incising into older braided paleochannels with climate amelioration at Pleistocene's end. Drawing by Patricia Demitroff.

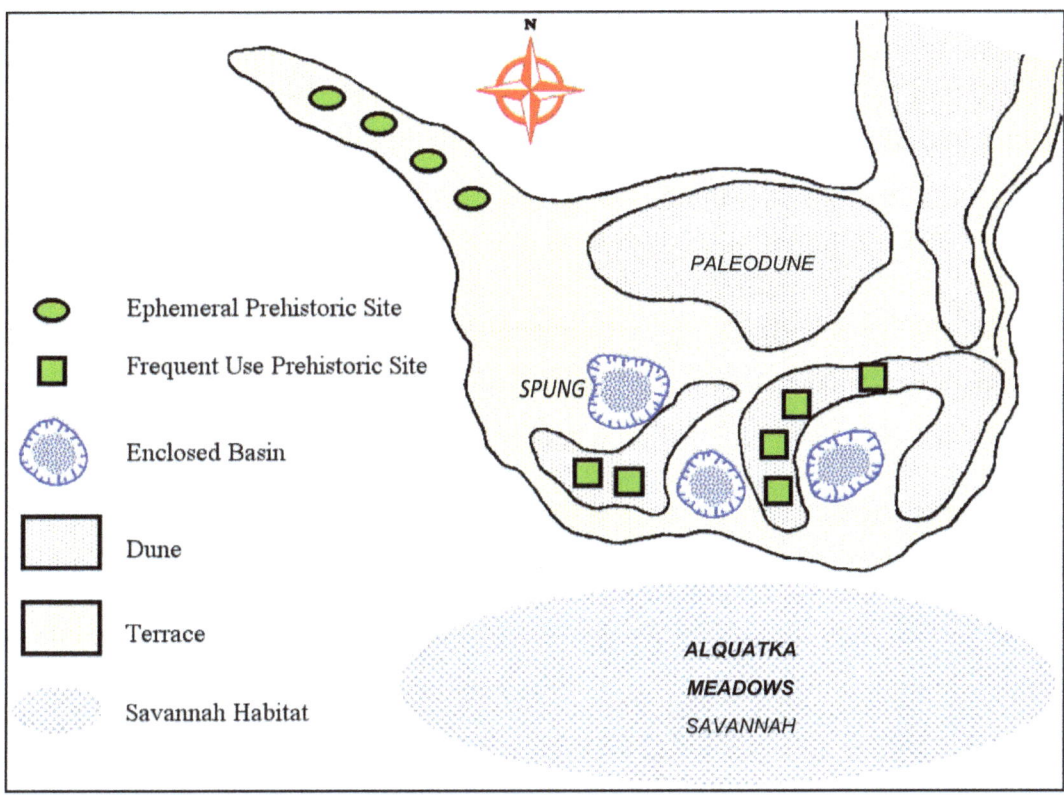

Fig. 7.11 Savannah topography and precontact cultural use.

Adapted from Cresson et al. (2006). (Fig. 7.9).

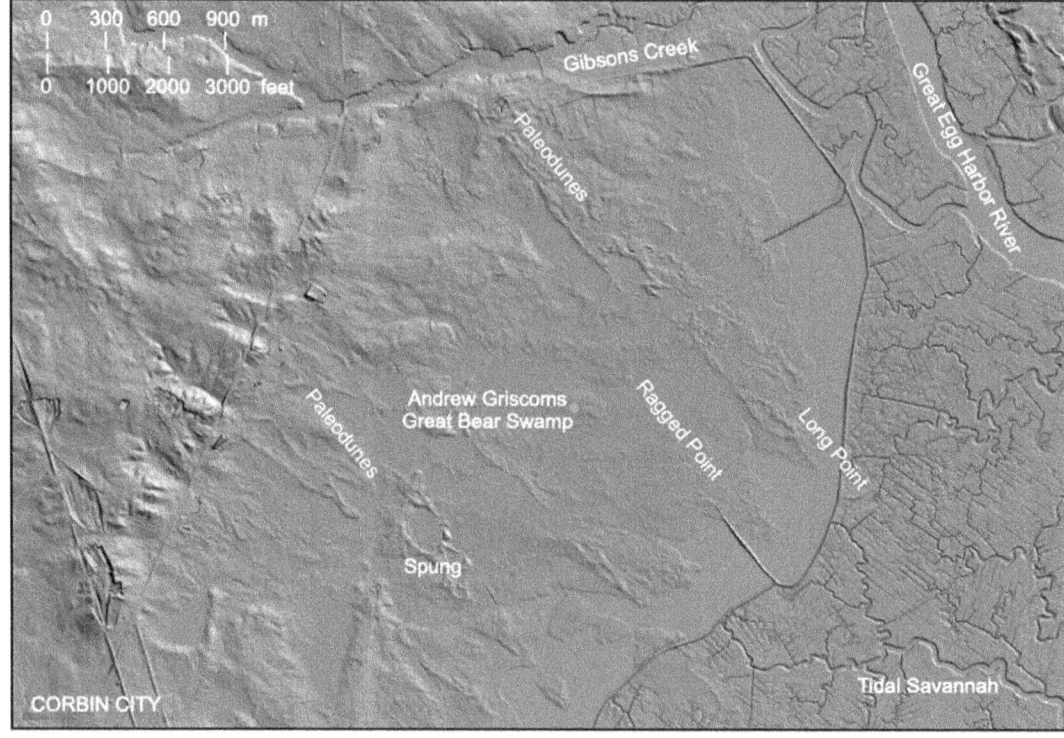

Fig. 7.12 Tidal savannah and cultural interaction with paleodunes.

Where dune arms cross ancient river terraces and continue onto tidal savannah flats, they form "points," their crests aid passage far out into meadowland. Not far from Ragged and Long Points is Job Somers Point, site of a 1695 ferry on the Long-a-Coming trail. Travelers used similar point dunes to cross the tidal marsh here to reach a store and ferry slip. The Estellville Glasswork, north of Gibson Creek, used the paleodunes as their sand source (Anonymous 1879). In application, sand quality is important. It must be free of silt and clay, which by nature dunes are due to wind sorting. Lidar imagery (NJOIT-OGIS 2019, accessed through Lidar in the Mid-Atlantic, Ostroff 2022).

Fig. 7.13 The early meaning of points and necks.

Apparently, there are two types of "points"—one from a nautical perspective (1) and another from a terrestrial perspective (2). Above, Doctors Point and Deep Point on the Mullica River are examples of the former sort (1), present at river bends. Leeds Point and the aforementioned Long and the Ragged Points (Fig. 7.12) are examples of the latter sort. According to OED, the chiefly US extended use describes "a projecting extremity of a wood, forest, etc." These are often the wooded hummocks one sees rising above the savannah flats.

Neck of the woods originally referred to a narrow stretch of wood or pasture, and later to a wooded remote settlement (OED 2003). In the Pine Barrens sense a neck (3) is often peninsula-like upland ground that is found along a river and bounded by water or wetlands (see Marsh et al. 2019). Rivers were the highways and a position in association with them was important to note. The author's experience is that "neck" (e.g., Broad Neck, Great Neck, Tar Kiln Neck) is used early in regional settlement at a time when rivers are more dependable transportation corridors than early trails.

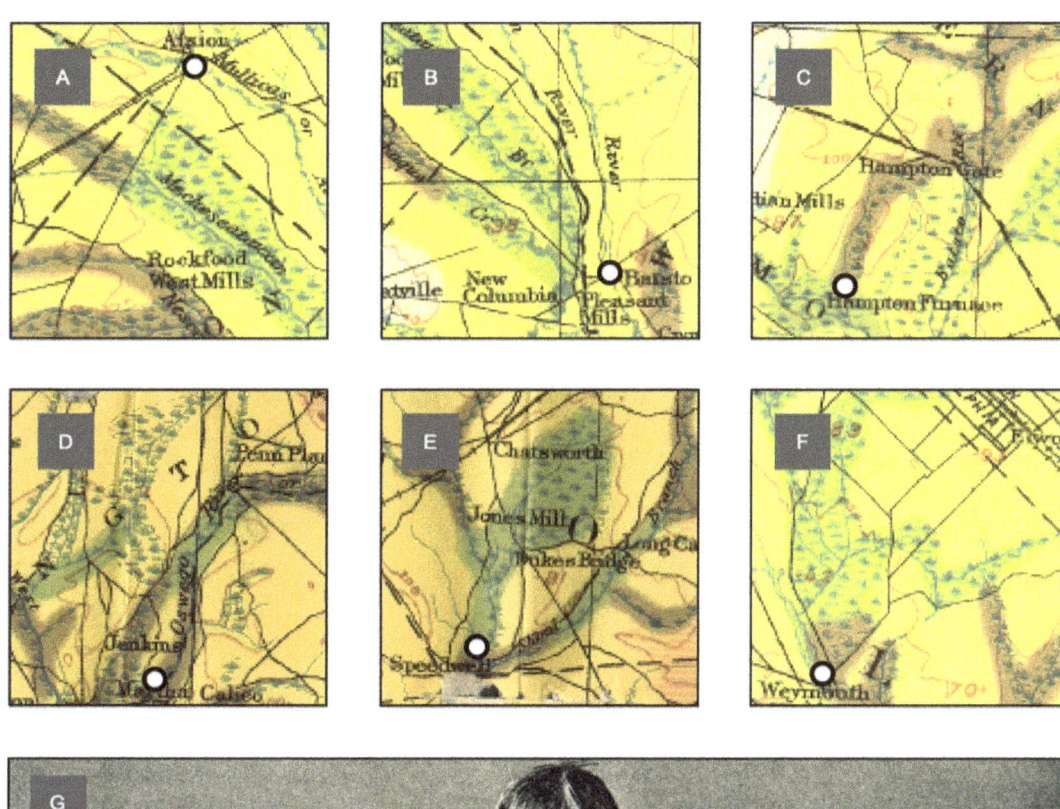

Fig. 7.14 The relationship between savannah and iron production.

(A–F) Bullseyes mark the locations of bog-iron furnace operations. The Figure is adapted from Harshberger (1916, Plate 1), with proximal savannah habitat shaded in green. The tree-sparse meadows of savannah would have been a convenient location for bog-ore extraction.

(G) Raising ore at Martha Furnace. Copy of a glass lantern slide. The illustration is signed Kenneth W. Thompson and based on Moore (1943). The illustration was published in Baer (1944). Reproduced with permission from Hagley Museum and Library, Clement S. Brinton collection on the early iron industry.

Fig, 7.15 Pine plains are among the most distinctive of Pinelands landscapes.

A narrow trail climbs to the top of the plateau just above Watering Place Pond (Harshberger 1916: 143–144). Near Warren Grove, NJ. Photo by Susan Allen (Picture Stockton . . . from the viewpoint of Apple Pie Hill Fire Tower, December 14, 2018, https://stockton.edu/news/2018/picture-stockton-from-apple-pie-hill.html).

Fig. 7.16 Pine Plains were once present in Cape May County.

Belleplain, a hamlet by Woodbine, NJ, was once a pine plain. Map portion reproduced from Gordon (1850).

Fig. 7.17 East, West, and other pine plains.

Even into the turn of the twentieth century, large swaths of the Pine Barrens were deforested. Engle et al. (1921: Plate II) documents pine plain around Millville, NJ. Little has been recorded on pine plains beyond the extant of the East and West Plains. Such habitat once supported a now-extinct species of heath hen.

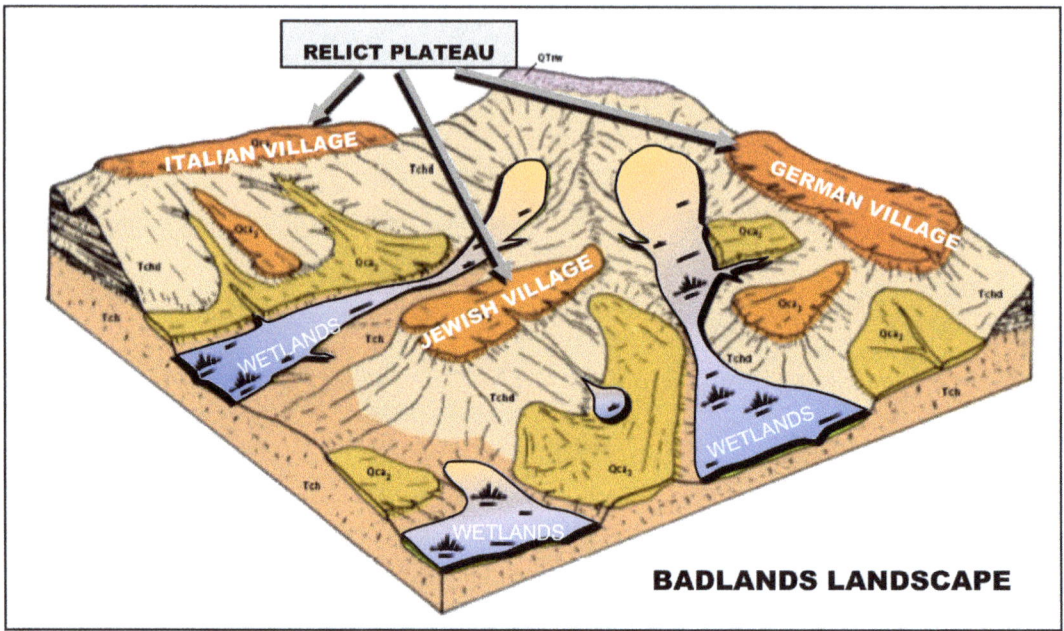

Fig. 7.18 South Jersey's badlands of inverted topography, a wasting (antisyngenetic) landscape.

A block diagram showing schematic relations between Tertiary and early Quaternary deposits. During the Miocene a large river (Proto-Hudson?) emplaced the elevated surfaces. A long period of slow weathering followed deposition. Cold, dry, and windy conditions (i.e., periglacial) further modified this landscape during much of the Late Pleistocene (Owens et al. 1983; Newell et al. 2000; French et al, 2003). The younger slope deposits below are attributed to periglacial mass wastage of the Bridgeton and Cohansey Formation (Newell 2005; French et al. 2007). The diagram is adapted from Newell et al. (2000, Fig. 4). Marsh et al. (2019) suggest a model of railroad-era settlement where this badlands landscape templated an ethnic archipelago. Settlements were founded upon islands of relict plateau upland. Surrounding wetlands and waterways isolated the villages from each other, helping to preserve distinctive traditional neighborhoods.

08
Windows to the Ground Water

A Dwindling Natural Resource

Many elderly residents of the Pinelands tell stories of how the water has disappeared from cripples, swamps, streams, and spungs over their lifetime. The author has interviewed numerous old-timers in the pursuit of preserving the local lore. The locals assumed that the wetlands would naturally dry up through time. The classic model of bog succession states that bog communities will infill over time and climax with forest growth (Shimwell 1971; van der Valk 1981; Tiner 1985: 30, Fig. 12; Ricklefs 1990; Roberts et al. 1993; Smith 2012). However, careful examination of the literature that supports this model indicates that closed basin bogs should persist for thousands of years without disturbance (Klinger 1996; Middleton 2002, 2004). This chapter provides supporting observational evidence to help support or refute a future model-based study that looks into this purported loss of shallow groundwater, a narrative that is indelibly etched into the local traditions and knowledge base of those who work and live in the Pine Barrens. A recurrent theme of wetland degradation in collective wisdom, if true, has many implications for the overall well-being of the area's flora and fauna.

Accounts of the region's desiccation go back to the time of Peter Kalm, the botanist. Upon his visit to Raccoon (Swedesboro) in November 1748, he recorded some accounts of smaller lakes, brooks, springs, and rivers drying up over a period of many years (Kalm 1770: 185–187). The flow of larger rivers was reported to have remained about the same. To Mickle (1845: 150), "the fall of the waters, not only of our inland ponds, but of the creeks [and of course the river with which they connect] is a well established phenomena." In nearby Pennsylvania, the disappearance of creeks and "the diminution of waters [generally], which formerly abounded in this part of America" (Brissot de Warville 1792: 341) was attributed to deforestation. Abbott (1907: 62–65) noted the drying of *brook streams* over centurial to millennial timescale, likely caused by deforestation, land drainage and field cultivation. Of the region, Prowell (1886: 1) states, "The streams running out of the hills are rapid, yet the volume of water has materially diminished by the gradual removal of timber from the upland and swamps."

Farr (2002: ii–iii), in *Waterways of Camden County*, comments that "almost all streams have suffered a diminution in flow and size" in the region, and "many small streams . . . have disappeared" altogether. Accompanying the drought of 2000–2002, over a dozen newspaper and magazine articles addressed the apparent problem of South Jersey's dwindling hydrologic resources (e.g., Kaskey 2000; Hajna 2002; Moore 2002b). Fry (1916: 30), Landis' lead surveyor and local historian of the Vineland tract, noted the that "there was much more water in the streams and swamps in those days than now [i.e., 1916 vs. 1863]." In part this phenomenon might be associated with the collection of wetlands organics to "redeem our swamp lands and open all the muck beds" (Landis 1916: 41) for the public good to amend poor sandy soils. Able (2020: 47–48) in a monograph on the Mullica River valley suggests over-withdrawal has occurred. Early on, the Pinelands Commission did not view the issue of the ponds drying up as a problem (Statewide Watershed Management Advisory Subcommittee, Meeting Highlights, December 12, 2000).

Others have not seen a diminution in streamflow or groundwater levels, at least as of the turn of the twentieth century. Gifford (1900) bucked the popular notion of his time that streamflow was declining. On the contrary, he stated that "there is no competent evidence, based on careful measurement, that there has been any important change in the streams, except what may be accounted for by artificial diversions of their waters, or by the fact shown by rainfall records that protracted droughts have been more frequent during the last quarter-century than were during the previous quarter" (Gifford 1900: 132). This was verified by interviews of many old mill owners who invariably claimed, "My stream is better now than it was when I was a boy" (Gifford 1900: 161–162. In the management plan for Atlantic County, it is stated, "there is plenty of water in the major aquifers (Dovey 1977: 29). Creek dredging by federal workers to draw off "the stagnant water" raised concerns that water drainage would become too rapid and desiccate the surrounding Barrens

(Chalmers 1951: 140). Mosquito ditching destroyed ponds in Cape May County (McConnell 2014: 308–309), once valuable bird habitat (Wichansky et al. 2006: 138) compared 1880s and 1990s maps and concluded South Jersey's wetlands have changed little in the interim, attributing this to New Jersey's strict environmental laws.

There is a shift from concerns solely on Pinelands water quality (Bizub 2008) to that of water quantity too (Lacombe et al. 2009; Bizub 2010; 2014a; Somber et al. 2013; Kecskes 2014; Conegno 2014; Post 2014; Stutz 2017). There has been some questioning of the reliability of hydrological modeling (Kavetski & Clark 2010). Across South Jersey Konikow (2013) modeled a groundwater depletive use of 1.3 cubic kilometers (0.3 cubic miles), or ~ 1,075,100,000,000 liters (~284,000,000,000 gallons), although the metric does not specifically tweak out a Pine Barrens over-withdrawal. Importantly, the shallow, near-surface groundwater supply that sustains wetlands is dependent upon the thinnest skin of water perched atop a massive aquifer (Roman & Good 1983), which in practice confounds a computation's ability to detect such a minuscule loss.

Accurate records of shallow aquifer levels in the Pine Barrens are scanty, episodic, and of insufficient duration to document long term trends in ground water base level (Epstein 2003). Preliminarily, it is suspected that over-withdrawal from the Kirkwood-Cohansey aquifer by the Atlantic City region (Fig. 8.1) is depleting streamflow and ground water across watershed boundaries (NJDEP 2003a: 54). In the Mullica River basin, a historic record of head decline of ~1.0 to 1.5 m (3 to 5 feet) had occurred between the late nineteenth and mid twentieth centuries (Rhodehamel 1973: 24). In the absence of long-term ground water data, an account of the physical condition (e.g., depth of water-fill, duration of water-fill, complete absence), spungs, cripples, blue holes, and savannahs can be used to indicate historical changes in ground water amount. In this chapter, they are used as "windows" into the Pine Barrens water table.

Spungs as Indicators of Hydrologic Change

An important reason why many of South Jersey's spungs have been overlooked is that they remain waterless for longer and longer periods (i.e., shortened hydroperiod), and many have dried completely.[1] Five large ponds, linked by the Egg Harbor trail system, are used as examples to illus-

---

1. e.g., Farr (2002: 14, Big Spung; 111, Moss Pond).

trate this point. Each of these watering holes was once a well-recognized landmark by seventeenth-, eighteenth-, and early nineteenth-century travelers. A colonial period traveler from the Delaware River would use this trail network on his way to Egg Harbor, then the name for eastern Atlantic County, and possibly also the name of a village now lost to history (Mickle 1845).[2] Nearly all of these enclosed wetlands are unrecognizable as ponds today.

Broad Pond

A journey along one of the pond-linked Egg Harbor trails is used to illustrate the disappearance of spungs from the landscape. We begin with the Broad Pond (Figs. 4.6D, 4.22, 8.2), a mega-spung that once was one of the largest water bodies in Salem County. It is positioned between Elmer and Willow Grove, in the Pine Barrens section of Pittsgrove Township. The author learned of this site when hired to evaluate some trees for the Broad Pond Refuge, a private animal preserve. The owners assumed that Broad Pond was the small, recently excavated fishpond next to the preserve residence.

During research on the provenance of its name, it came to light that Broad Pond had been a prominent feature on many old maps and surveys (Nelson 1781; Everts & Stewart 1876; Woodward 1964: 8–9). The Egg Harbor Road (Nelson 1781) passed the southeast corner of this spung, on its way through the woods toward North Vineland. The region southeast of Broad Pond was called "Broad Neck," deemed a desolate barrens whose "inhabitants were not noted for their enterprise, or for being very unexceptional citizens" (Cushing & Sheppard 1883: 465). Broad Pond is, today, a low-lying wood, and has disappeared even from the memories of Pittsgrove Township's oldest residents.

---

2. Mickle (1845: 111–112) wrote: "On Great Egg Harbor Bay, in Gloucester county, according to Scott, there was formerly a town called Egg Harbor, the inhabitants of which exported large quantities of pine. As this writer lived in Philadelphia, and compiled his work with a great deal of care, we have no doubt there was a village of this name; but where it stood or into what it has been changed we are unable to tell. Such, such are the works of time!"

Egg Harbor Pond

Continuing eastward 13 km (8 miles) along this Egg Harbor trail by way of the Forked Bridge[3] (*Willow Grove* [Hartman 1978: Map 5]) over the Maurice River and heading southeast partly along the Mollago (Malaga) trail (Purvis 1903) and back on the Egg Harbor trail, is the next major spung, named "Egg Harbor Pond" (Figs. 4.17C, 8.3) at the original site of the original Marcacci's Meat Market in Vineland. The Hollingshead Survey (Hackney 1795) of 7,941 hectare (19,623 acres), drawn from the Penn's survey, clearly marks the location of the pond along the Maul's Bridge section of the Egg Harbor Road (Hampton 1917). From the ground, there is nothing to indicate that this once contained a great body of water. However, from an oblique air photo taken during the wet winter of 2001, its parabolic outline is easily discerned in the open farm field (French & Demitroff 2001: 340, Fig. 2A). Many spungs probably share Egg Harbor Pond's rounded saucer-like shape, except their margins are now often obscured by thick vegetation. The distinctive parabolic dish form seen on laser altimetry helps to identify other potential spung sites that have long since dried up as well along the Egg Harbor way (e.g., basin features at heads of Green and Endless Branches).

A local farmer (personal communication, John Marcacci, 2001), now deceased, recounted his childhood memories of Egg Harbor Pond, when water still filled it year-round. He stated that in the late 1920s the pond had sufficient quantities of water to support fishing, boating, and swimming. In the spring, its margins would reach almost to Vine Road, several meters (yards) higher than the ancient pond bottom. Vine Road had to end its westward trek at the Central Jersey railroad track, since the terrain was too wet to continue on to Main Road. Now Vine Road continues to Main Road over dry ground. Powerful springs fed Egg Harbor Pond, which the Vineland Railway tried to cap, unsuccessfully, by dynamiting them around 1870. By 1960, the pond had lost so much water that a slough was dug for

---

3. Alternatively, a wayfarer on Egg Harbor trail once across the Maurice River might take the northeast route of the Forked Bridge trail, a spur of the old Woodbury trail. Within six km (four miles) the traveler would pass The Park savannah to a great circular spung called The Lake (Fig. 4.5B), then cross the c. 1761 Tuckahoe trail at Miller's Place (Clement 1845: 71, later Blue Bell Tavern), enter the Newtonville dune field (Demitroff 2007) to cross wetlands, head Indian Branch, and eventually meet up with the Long-a-Coming trail at c. 1738 Smith's Little Mill (Steelman 2003) above Weymouth before entering the Lochs-of-the-Swamp. The Lake spung was drained for farming c. 1890 by Italian farmers as it was already drying up (Dillingham 1911: 88–89).

irrigation purposes in its center (French & Demitroff 2001: 340, Fig. 2A), as an attempt to reach the dropping water table.[4]

Spung at Campbella

Another spung (Fig. 8.4) was present at the Buena intersection of US 40 and NJ 54, 6.5 km (4 miles) east of Egg Harbor Pond. Three ancient trails met at this location, the Cohansey (an Egg Harbor), the Blue Anchor (Stewart 1932: 35, called the Shamong trail), and the Tuckahoe (Hackney 1795; Hartman 1978: Map 5). In 1767, a plantation known as Parson's (Parsons Tavern?) began here (personal communication, Glenn Bingham, Mark Parsons, Joan Best, August 6, 2018). Unsubstantiated lore places a Campbell's tavern here too in 1773 (Hutchings 1907b); records found by Bingham, Parsons, and Best indicate Veals—not Campbells—founded this establishment. The hamlet grew in support production of naval stores for the shipbuilding trade (Bauman & Cassaday 1999: 6). A powder-charcoal mill was built here just after the War of 1812 (Hutchings 1907a). Its three-ton cast iron crusher wheel still marks the mill location. A small stream issued from the Campbella spung, and passed eastward towards Deep Run (personal communication, Stanley Reed, 2000). The wooden bridge at this location on the Blue Anchor Road is long gone, as is the spung and its run outflow drainageway. A Wawa convenience store has recently been built directly on the old pond site, despite rules in the Pinelands Comprehensive Management Plan that prohibit the action (Harper 2004).

Horse Break Pond

The fourth spung, Horse Break Pond (Figs. 4.3, 8.5) in Buena, is a little over 1.5 km (1 mile) away on the Egg Harbor trail. On a survey for a court case (Wright 1867a), the Egg Harbor Road (Beers 1872b) again strikes the southeastern corner of a spung, heading Horse Break Branch. The site is purported to be adjacent to an early Swedish settlement (personal communication, Stanley Reed, 2000), complete with a burying ground.

---

4. Scagnelli and Moore (2009: 63) report that historically sixteen rare plants were reported from this site, of which only two remain—found in the slough. Two additional rare species were found here that were not reported on earlier records.

A bouncer at the Midway (Veal, Campbell) Tavern is reputed to have removed the gravestones for use in building a walk during the 1940s.

South Jersey's trails have long been associated with illicit affairs. Smugglers favored these twisted cartways to avoid British customs and tariffs in colonial times (Pierce 1960). During the nationwide Prohibition (of alcohol) from 1920–1933, illegal stills were commonplace in the Pines. The myriad of trails provided easy escape for "rumrunners" in their pursuit by "revenuers."

It is the author's belief that there is a significant relationship between these trails and the Underground Railway. In the mid-1800s, a charcoal camp known as "Stephan Colwell's Coal Grounds" (Beers 1872b) developed adjacent to Horse Break Pond. The nearby Union AME Church, a log structure, was built in 1858 as a sister congregation to the Springtown AME in Greenwich (Petway 1961; Steelman 1965: 31; Barrale 1979: 21–22; Trusty 1997: 251). The latter church is linked to the former by the ancient Cohansey trail and served as refuge for runaway slaves (Trusty 1997: 72–85; National Park Service 2018).

The Black abolitionist William Still was a prominent patron who financed escape networks throughout the Delaware Valley. He made his fortune through the charcoal trade in Philadelphia as a coal porter (OED, September 2013: "A person employed in carrying of unloading coal"). As a young man, he, and his brother James Still, the famed "Black Doctor of the Pines," cut cordwood for charcoal at the Brotherton Indian Reservation (Khan 1872; Still 1877) on the Shamong/Blue Anchor trail (Stewart 1932: 35). The Still family remains well represented in the area today. This may have been part of the economic engine that drove the Underground Railway and warrants further exploration. Later, abandoned charcoal camps become settlements for Blacks, Jews, and Italians (e.g., the New Rome section of Richland—Black to Italian to Black; Mizpah, Bears Head; Woodbine—Black to Jewish to Black).

Like its Swedish settlement and charcoal camp, Horse Break Pond itself is largely forgotten, as is its now mostly dry Horse Pond Branch. The meadow marking the spot has a shallow fill of water every now and then. But this is not quantity enough, nor is it filled often enough, to slow encroachment by trees. The pond is not expiring from infilling but is simply starved of groundwater. Adjacent savannah at the Townsend Swamp is now dry ground (Fig. 8.9).

Lochs-of-the-Swamp

The trail then continues 13 km (8 miles) to a ford across the Great Egg Harbor River (Fig. 5.4, NE of cripple), passes the Dead River (Fig. 5.11B) whereupon circuitously it meets with three large ponds[5] (Figs. 4.10, 8.6) and numerous smaller ponds at the Lochs-of-the-Swamp (Denny 1774; Wright? c.1867; Wright 1868; Blackman 1880: 408–410). Lookout Pond (Figs. 4.9C, 4.10) was over 1.5 km (1 mile) in length. Brake and Crane (Figs. 4.9C, 4.10) Ponds were about half the length of the first pond. Their demise may have been due at least in part, to the exploits of Charles Makepeace (Blackman 1880: 416; Cain 1958: 35), in his quest to develop cranberry bogs in this area. From this point, a traveler could take the Long-a-Coming trail (Chalmers 1951, also see Stewart 1932: 35, branched with Maeroahkong trail) southeastward or northwestward as it linked a chain of spungs, blue holes, and savannahs. Another trail led to the charcoaling settlement of Carmantown (Beers 1872a; 1872b; Fritz 2012), settled around another cluster of ponds such as the Peachy Pond (personal communication, Elmer Riply, 2017), the Little Goose Pond, and the Big Goose Pond (Fig. 4.8C). Heading due north on the Kings Road (Wright 1868; also see Stewart 1932: 35, Maeroahkong trail) brought one to the geologist Wolfe's hometown of Nescochague. Here the trail met up by Hammonton Creek with Old Forks Road, an early spur of the Long-a-Coming trail that split off at the Beebe place and went to The Forks-of-the-Mullica (Clement 1845: 70). This village was likely titled for the wetlands[6] present, as this aboriginal name roughly translates to "wet, grassy, and muddy" (Becker 1964: 48; also see Grumet 2014: 73). Scots settled Nesco at the site of a Lenape camp in 1707 (Green c. 1920s?: 4–5), and its residents may be responsible for designating the ponds as "lochs"[7] down-trail.

---

5. The three ponds on Three Pond Branch along the Blue Anchor trail are often confused with the adjacent all-but-forgotten Great(e) Ponds (Clement 1852: 69; Hopkins 1873: 23), which have disappeared from all living memory. They resided in the paleochannel west of the polygons, which was later sapped of its water (beheaded) by a small branch of Three Pond Branch. The Branch's true location can be seen on Fig. 6.11B. Like the Great Ponds, the true Three Pond Branch has all but faded away.

6. In 1761 there was a Long Meadow Branch (Clement 1845: 70), suggesting savannah habitat once existed around Hammonton (Bishop 1831) and still exists along the nearby Nescochague Creek (Smith 2012: 7). Spungs are also present, like Desolation Pond (Clement 1845: 58) on Old Forks Road five km (three miles) west of Nesco.

7. Locks Bridge on the Mullica (three miles below Atsion) apparently had a sluice or gate to allow river traffic (Royal Colony of New Jersey 1765). Locks Branch (Clement

Numerous stories are told of fish having been present in spungs such as Grassy Pond (personal communication, Edith Fanucci, September 1999, see Harshberger 1916: 143) (Fig. 4.14), Egg Harbor Pond (John Marcacci, personal communication, 2001), and a now-waterless spung at the head of the Manumuskin River's East Branch near the Punch Bowl (personal communication, Eric Hensel, Sr., July 2004). Another spung with a fish population is Seaf Weeks Pond[8] above Lower Bank (personal communication, Carl Farrell, July 29, 2018). At the turn of the nineteenth century, a biologist (Cope 1896) was mystified by the number of fish he found in the shallow depressions (spungs) of Camden County's Pine Barrens. The pools were small 7.5–9 m wide (25–30 feet), shallow (>1 m [3 feet] deep), and not in communication with any larger water bodies. Yet many of the fish were fully mature. It is likely that other spungs once held water long enough to support fish.

Cripples as Indicators of Hydrologic Change

Cripples also provide clues about the health of an area's groundwater. Surface runoff is negligible in the Pine Barrens. A cripple depends upon shallow groundwater for recharge, as do Pine Barrens' rivers and spungs. If ephemeral flow through these waterways diminishes or disappears altogether, a drop in the local shallow-water table is the likely cause.

Settlers arrived at Union (the Union House site) on the Cape Road, which was laid out along an Indigenous trail prior to 1729 (Elmer 1869: 73). A nearby cripple had an interesting metamorphosis of name (personal communication, Bob Francois, August 14, 2018). Francois's grandfather[9] told him of how the Cape Road forded a small, perpetually wet valley just to the east, and so it was called "Horse Wallow." Older maps of Cumberland County, New Jersey, distinctly indicate water in Horse Wallow (Beers et al. 1862; Stewart 1876: 6). Over the years, the spot became dryer and

---

1845: 71; 1867: 25, Lake Branch with Israel Locke residing by it) is named after a Swedish settler although there are loch-like ponds nearby (e.g., the Great Ponds). Lock Branch a km (half mile) above the Little and Big Goose Ponds appears to have once supported a loch-like pond on c. 1931 aerial photomosaics. A loch, a lock, and a Locke; such are the complexities of place names.

8. Farrell recites duck hunting here around 1978 with Lower Bank resident Bruce Wise, and a wind gust blew his sandwich into the spung—only to be quickly eaten up to his surprise by fish.

9. Albert Beebe.

dryer. By 1890 crossing became known as "Horse Hollow" since it was no longer wet. It remains dry today (Fig. 8.7).

As with Horse Hollow cripple, the Emmelville cripple (Figs. 5.1, 5.4, 5.11) no longer behaves as an intermittent watercourse. According to a local game warden, the latter consistently had sufficient winter ground water replenishment to allow winter skating at the location shown in Fig. 5.1. His family farmed this location for several generations. Artesian springs, now gone, once filled spungs up-valley from the Emmelville cripple. Wood Ducks once nested in the cripple valley just above its junction with the Great Egg Harbor River, but no longer do (personal communication, Austin Pirrone, September 16, 2006).

Springs as Indicators of Hydrologic Change

Many springs—confined[10] and unconfined[11]—have disappeared over the last century. Pamphylia Spring, on the banks of the Cohansey River in Bridgeton, was once so mighty that in 1716 Pamphylia Township was proposed a namesake in its honor (Heston 1924: 821; Elmer 1869: 30, 48) as springs generally had important cultural significance (Klempe 2015). A local archeologist is the only individual this author has located who remembers this spring in an active state. During the 1930s, fisherman could still fill their jugs with its waters to slake their thirst during a hard day's outing. It had a modest flow in the 1930s, and the high tide would cover the source. Still, it would bubble out of a small barrel sunk into the mudflats halfway between the high and low tide mark (personal communications, Jim Holder (1928–2011)—first-hand accounts, January 2003; Jim Parkhurst (1939–2015)—second-hand accounts, March 2004). Its whereabouts are not readily apparent today.

The artesian "boiling springs" of the Manumuskin River had wide recognition in the nineteenth century when Bowen (1885: 39) wrote his *History of Port Elizabeth*. The sparkling and hygienic water bubbled profusely from the riverbanks. Despite the relatively pristine state of the river's surroundings today, the springs are gone. Gone too are the boiling springs (Rose et al. 1878: 21) that made salt savannah pasturable

---

10. Confined aquifers have an impervious bed above and below them.

11. Unconfined aquifers have an impervious bed beneath them and the water table, aeration zone, and surface above them.

and provided valuable bird and fish habitat.[12]

When Chalmers (1951: 84–86) played on the banks of the Great Egg Harbor River as a child of the 1880s, she recounted the Inskeeps Blue Hole[13] as a fantastic feature with abundant springs. Its strong flow would create a whirl-pooling effect, adding to its mystery. The folklorist reported in the 1930s that the hole was still sinister but was probably larger some years ago (Beck 1937: 148–153). Today, it is a shallow pool, incompletely filled, and with only an occasional hint of its former blue tint. Like the Dog Heaven in Farmington (Anderson: 1964: 68), the Danger Hole (Fig. 8.8) by the author's home along the South River, and many other nameless blue holes throughout the Pines, its artesian head has long-since faded.

Savannahs as Indicators of Hydrologic Change

Little (1946: 37, 39) theorized that wetland meadows could have been formed in the Pine Barrens as a result of dry-season wildfires. If a forest fire consumed the wetland's trees and shrubs, but left the organics-rich surface mat unchanged, woody plant regrowth would have occurred rapidly. However, if the organics-rich surface layer was through combustion consumed to the water table, then the woody vegetation's regrowth would be delayed until the *Sphagnum* mosses accumulated in sufficient quantity to create a new seedbed. Woody vegetation could only germinate and survive in a substrate that was sufficiently well aerated to allow respiration (i.e., above the water table). (Little 1979: 308–310).

Whittaker (1979: 320) agreed that dry-season fires, and soil factors, probably created the savannah-like communities Harshberger described.

---

12. Price stated (Rose et al. 1878: 21): "In numerous places, all along the coast, and often hundreds of yards from the upland, may be found fresh-water springs bubbling up through the sods. The water is pure, fresh, and cold. They supply drinking-places for cattle pasturing on the salt grasses, when it is not too miry around them, and are the choice resorts of the poor harassed wild fowl, who seek them in the night. Springs of fresh water are also known to exist in the channels and bottoms of the creeks which flow through them. The continual boiling and bubbling of these through the muddy bottoms make deep holes in the creeks, which are the favorite abode of eels, perch, and other fish who love fresh water. These places are eagerly sought by the cunning angler, who is pretty sure to find his unsuspecting prey 'at home' in them."

13. Chalmer (1951: 122 & 123) makes reference to an old cow path to Cow Swimming in Folsom, another forgotten hole that has remained elusive to this author. Also forgotten is a nearby cluster of ponds on the Long-a-Coming trail west of Folsom, near Winslow, called "The Ponds" (Mattocks 1762).

A high water table is necessary to have savannah-like conditions (Smith 2012: 62). At the braided stream deposits near Atsion (Farrell et al. 1985), a severe fire in 1977 burned the covering vegetation and the organic mat down to the mineral soil. Eighteen months after the fire, pine seedlings sprouted everywhere, except in the lowest troughs. Today, the pitch pine forest reaches +10 meters (+33 feet) in height at Farrell's research site. Meadow-like conditions are limited to the interstice formed by the lowest channels between the ancient bar deposits indicative of raging floods like the bar system he mapped in Alaska on the Copper River delta or in front of the Malispina Glacier (personal communication, Stewart Farrell, August 13, 2018). Harshberger (1916, Plate I), and Cook and Vermeule (1890, sheet 15), indicate a more substantial savannah once existed at this spot.

If proven a widespread practice, cattle grazing could have adversely impacted savannahs early on. Recent investigations (Middleton 2002; 2004) have shown that moderate to heavy grazing of meadows damages sedges through trampling of the tussocks, allowing the invasion of woody plants. Because use of savannah as a biome for pasture was a rarity in the Pine Barrens by the late 1800s (Gifford 1900: 248), hence it is unlikely Harshberger's (1916) meadows were significantly affected by cattle grazing. According to Gifford (1900: 248), constant burning for berry production and occasional grazing could gradually kill weakened grasses and deteriorate a savannah habitat.

In Little's (1979) model of meadow formation, trees should not have reestablished themselves at the Atsion site until a layer of *Sphagnum* mosses accumulated. Instead, the area quickly reforested, despite an insufficient amount of time for the mosses to have amassed. The mineral soil remained dry enough to promote germination of woody plants. Mulhouse et al. (2005: 417, SC), Chmielewski (1996, SC) and Laidig (2012, NJ) found that grass/sedge dominated enclosed basins had shorter and less variable hydroperiods than ponded and meadow dominated enclosed basins, suggesting savannah habitat requires a relatively reliable water source. Perhaps the disappearance of savannah vegetation at the Farrell site, as well as at other savannah sites throughout southern New Jersey (e.g., Fig. 8.9, Townsend Swamp savannah) could be explained by a recent lowering of this water table. Other factors that might help explain a reduced savannah hydroperiod include: 1) an increased surface runoff within the hydrologic system; and 2) an increase in the evapotranspiration rate (e.g. more or denser vegetation cover, longer plant-growth periods) would equate to a diminution in savannah groundwater (personal communication, Claude Epstein, October 12, 2018).

204   SOGGY GROUND

Fig. 8.1 Over-withdrawal of the 800-Foot Sands.

A well-drilling rig in Atlantic City, providing water from the 800-foot sand unit of the Kirkwood-Cohansey aquifer. The water level in the once artesian aquifer has declined one to two feet annually since the 1890s. Water is being used in an unsustainable way. (The NJDEP (2003b: 7) warns, "Pumping from the Atlantic City 800-Foot Sands is also likely to impact surface water base flows particularly over the long term in the western portion of the county." NJDEP (2003b: 9) further states, "A comparison of the water supply and existing water demand estimates indicates that both the Great Egg Harbor River and Mullica River watersheds are already overdrawn." Has the anticipated drop in baseline already occurred? Photograph is from Thompson (1928).

Fig. 8.2 Broad Pond's current state.

The once famed spung has changed into a damp hardwood forest managed by the Broad Pond Refuge, a private wildlife conservancy. The entire right-of-way, from immediate foreground (east) to the clearing in far background (west), transects the now-dry spung bottom. Near Elmer, NJ.

Fig. 8.3 Egg Harbor Pond's current state.

This famed pond (saucer-shaped depression in upper center of photo) was an important waystop along the Cohansey-Egg Harbor trail. From the ground, there is little to indicate that this location once contained a great body of water. However, from this oblique air photo, its parabolic outline is easily discerned in the open farm field. Until the 1920s, this pond was swimmable, boatable, and fishable. By 1960, the water table had dropped precipitously. The property owner dug the long slough in the center to again reach groundwater for irrigation (personal communication, John Marcacci, June 2000). Photo taken by Ron Bertonazzi and Mark Demitroff. Vineland, NJ.

Fig. 8.4 Campbella Pond's current state.

This pond was at the intersection of the Blue Anchor, Cohansey, and Tuckahoe trails. It is the location of an early tavern and its associated Buena Powder Charcoal Mill (c. 1815). The spung escapes modern perception since it only rarely fills today. In 2004 a Wawa food market was located on the southern rim (center-right of photo). In has since expanded by filling in the wetland, despite stringent Pinelands zoning laws in place that normally protect such an ecological, archeological, and cultural resource (Harper 2004; NJ Pinelands Commission 2015a). Photograph taken during the spring of 2002. Buena, NJ.

Fig. 8.5 Horse Break Pond's current state.

This pond on occasion has water within its basin for several months. Woody plants are encroaching from the edge as the pond fills less frequently. New Rome, near Buena, NJ.

Fig. 8.6 The Lochs-of-the Swamp's current state.

Built land between Lookout and Crane Ponds along US Route 322 at the Lochs-of-the-Swamp, both likely drained by C. Makepeace. Considered "harbor for vermin and breeding places for mosquitoes" (Blackman 1880: 419), some spungs were converted into cranberry bogs by Makepeace (Johnson 2001: 104). A former Makepeace bog developed at the Lochs' is now a blueberry farm (*left*). Crane Pond is now a mixed hardwood/pitch pine lowland (*right*). Great Neck, Mays Landing, NJ.

Fig. 8.7 Horse Wallow's current state.

A portion of the Hance Bridge trail (berm left & right of shovel) where it crosses Horse Hollow (*swale perpendicular to berm*) illustrates the historical drying-up of a round-bottomed cripple in an ancient dune field. Although a dry abandoned valley today, during the nineteenth century this location was wet. Horses mired here, accounting for its original appellation of Horse Wallow. Union, near Millville, NJ.

212   Soggy Ground

Fig. 8.8 Danger Hole's current state.

For years, a metal sign warned bathers not to dive from the ironstone bridge over the South River into the Danger Hole (Fig. 6.7). Zambone, born nearby in 1901, recounted the drowning of a student who dove into the deep, clear spring-fed pool that was the impetus for the alert (personal communication, Frances Zambone, 1977). Photograph taken by Joe Rosemont when the South River lowered and briefly dried for the first time in memory during the 1963 drought. Since the 1980s, this stream and its pair of riverine spungs called locally known as the Skating Ponds (Fig. 6.8) have become intermittent. This bridge was midway between the author's family farm in Buckhorn and the family mill in Richland, NJ.

Fig. 8.9 Townsends Swamp Savannah's current state.

This former meadow no longer behaves like a wetland, yet its tree component (e.g., pitch pine [*Pinus rigida* P. Mill.], black gum [*Nyssa sylvatica* Marsh.], American holly [*Ilex opaca* Ait.]) remains scattered and park-like. Bordering the Cohansey–Egg Harbor trail east of Horse Break Pond, Richland, NJ.

09
Spungs, Cripples, Blue Holes, and Savannahs

Ice Age Inheritance

A very real danger is that one can read too much into South Jersey's periglacial landscape, perhaps this author too. From 1945 to the 1970s, there was a robust movement (André 2003: 150, periglacial fever) to explain climatic landscape evolution across Europe. For example, freeze-thaw was a popular catchall term used to explain the formation of numerous phenomena. Recent advancements in geocryology (permafrost science) brought into question the role of freeze-thaw processes and other widely held views of traditional periglacial geomorphology (André 2003; French & Thorn 2006). Holocene overprint via physical, chemical, and biological processes has modified, and will continue to change the landforms of the Pine Barrens. The rigors of process-based scientific scrutiny cannot be ignored. Peers are encouraged to examine complementary material presented and use their discipline's scientific tools to bridge gaps beyond the scope of this monograph. Presented is a geographer's perspective, providing his interpretation of 200,000 years of natural history, and 14,000 years of cultural adaptation to the unusual environment of the Pine Barrens.

During the Late Pleistocene, southern New Jersey was at times a cold, arid, and windy plain. The geologic evidence suggests that climatic forces much different from those of the present reshaped the Pine Barrens landscape. In the High Arctic, periglacial landforms (e.g., patterned ground, mass-wasting features, thermokarst features) direct the size and shapes of wetlands today (Woo & Young 2006: 433). Similarly, ancient cold-climate conditions impacted the size and shapes of peculiar features found in the Pine Barrens today. The water table appears to have rebounded to somewhere near-present levels some 14,000 years ago, as a result of the moderation of climate and the associated sea level rise[1] (French & Demitroff 2001: 347). The rising water table provided a ready supply of water to the periglacial deflation basins and river channels, and abundant life returned to the Pinelands.[2] So began a near-continuous cultural pattern of resource exploitation in spung, cripple, blue hole, and savannah wetlands, although the Younger Dryas (12,700–11,700 years ago) record remains enigmatic.

First substantiated to appear were Paleoindians, who repeatedly visited "periglacial" features in a far-reaching pattern of hunt. Little is known of these scarce Late Wisconsinan associations, and their study will provide valuable insight into the earliest settlement of North America, Later Archaic and Woodland cultures used these features in a more sustained and diverse manner as part of localized seasonal rounds. Ice Age relicts such as basins, dunes, paleochannels, and springs were catchment areas for plants and animals important to foraging peoples. A network of interconnecting paths facilitated resource exploitation.

Early Europeans found advantages to adopting the Indigenous trails. They were exquisitely adapted over millennia to take advantage of natural resources. These tracks: 1) generally followed fast land (dry land above the high-water mark) especially along interfluve crests; 2) forded watercourses

---

1. During cold periods, sea level was much lower than today (Carlson et al. 2019). Terrain classified as Pine Barrens today became high and dry inhospitable ground. Pre-Clovis culture—if it ever existed—would have employed paleolevel springs (oases) that then emanated much further down the Continental Shelf in response to a lowering of regional groundwater—establishing a new piezometric baseline. Also, Ice Age spring-sites would have been closer to ancient lowstand shorelines with those upslope drying up due to a drop in hydrostatic head (Faure et al. 2002: 48, Fig. 1). Groundwater recharge would have been likely been restricted by a generally impermeable frozen ground for extended cold periods.

2. The Hackett mastodon displayed at the Rutgers Geology Museum was discovered in 1869 within a series of enclosed basins (Powley 1969: Sheet 22) at Mannington Meadow, Salem County (Salem County Tercentenary Committee 1964: 93). Silty Mattapex soils at this location were considered eolian deposits by Tedrow (1986: 300–301) and further interpreted as loess by Newell et al. (2000: Map D, Fig. 3) and Wah et al. (2018). An ancient trail at this location crossed Swedes Creek (Everts & Stewart 1876: 25) via Swedes Bridge. Swedish settlers were in the area as early as 1638 (Cushing & Sheppard 1883: 317). In Eastern Europe, mammoths would leave wet floodplains in summer and migrate to higher loess-covered interfluves where the ground was firm enough to support their great weight. Dry rims above thermokarst basins were productive summer habitat (Velichko & Zelikson 2005: 143).

where shoaled by transverse gravel bars or choked by windblown sands; and 3) linked productive hydrologic features like watering holes, springs, and meadows. Trail junctions at spungs were opportune locale for rest and watering livestock. Way-stops, taverns, and settlements sprang forth alongside water sources. Even today, South Jersey's roads approximate routes assigned by transhumance patterns long forgot, and await further study (e.g., Waldykowski & Krzemien 2013; Dalgaard et al. 2018).

Water Resources

The four hydrogeographic features chronicled here are literally windows to the shallow groundwater table. Their replenishment depends almost entirely upon seepage to a saturated (phreatic) zone above the uppermost confining layer. By observing how much water the features hold, and for how long they hold it, insights can be gained as to the region's overall hydrologic soundness. A long-term wetland baseline drop of only 10–20 cm (4–8 inches) "could cause significant alteration in [vegetational] community structure and composition" (Roman and Good 1983: 48, 'and habitat value,' 52; also see Staff of the New Jersey Pinelands Commission 1984: 6, "The ecological implications of removing even minor quantities of water from the Cohansey of the Pinelands are numerous" (also see McHorney & Neill 2007; Laidig 2012; et al. 2010; Zampella et al. 2010). Using a combination of historical records and anecdotal comments, it is possible to draw a reasonable conclusion as to the general trend of groundwater levels over time (Pastore et al. 2010; Haldersen & Treildel 2012; Jones 2013).

Concern over regional groundwater sustainability spurred an intensive combined investigation by the US Geological Survey, the Pinelands Commission, Rutgers University, US Fish and Wildlife Service, and the NJ Department of Environmental Protection. The combined studies advanced by the Gibson Bill (N.J.P.L. 2001 c.165) will assess current hydrologic conditions and develop a hydrologic database that includes predictive groundwater sustainability and flow models (Navoy 2004: 4–5). The Pinelands Commission approved the project in October 2003, and its completion was expected by 2009 (Zampella 2004: 1). The Kirkwood-Cohansey Work Plan included twelve[3] topics of interest. Those twelve

---

3. The work plan topics are: 1) hydrology; 2) wetland-forest community gradients; 3) Swamp Pink (*Helonias bullata* [L.]); 4) intermittent pond vegetation; 5) anuran-larval

studies were as a whole completed as of 2015 (Pinelands Commission 2015b). None of these directly address the historical context of Pinelands wetlands—Have the Pine Barrens been drying up? In the interim, a ground water depletion model was developed for the entire southern New Jersey region. Konikow (2013: 13, Fig. 7) predicted a ~1.3 cubic kilometer cumulative loss to that date. This is a "big picture" view of ground water depletion, which does not specifically account for the smaller watersheds[4] and even smaller sub-watersheds[5] (personal communications, Robert Kecskes, August 09, 2018).

When the first press reports (Fig. 9.1) of the long-term drying of South Jersey's wetlands appeared (Kaskey 2000; Hajna 2000a; Cresson 2000) state officials were unaware of the potential problem (Kaskey 2000; Hajna 2000a: 8A, 2000b), poor water-well documentation makes it difficult to document groundwater trends in a geological context. There are too few records available for study, they are of short duration, and they had not been uniformly logged (Epstein 2003). To some, this lack of adequate well documentation makes it impossible to scientifically prove protracted changes in the groundwater (Hajna 2004; Harper 2004). This view seems to dismiss possibilities afforded by interpretive local history and archeology, which has been added to "scientific reasoning" Frodeman (1995: 966). Scientific knowledge of the past is possible (Pastore et al. 2010; Haldorsen & Treidel 2012; Jones 2013), even if the conclusions are considered tentative (Gould 1986; Cooper 2002).

It is this author's conclusion that the water table throughout South Jersey has slowly declined for the past 250 years. This loss of groundwater accelerated since the 1890s. The exact reason for this decline is beyond the purview of this monograph and the author's skill sets with expertise in land-surface processes, not hydrology. Groundwater over-withdrawal is part of the problem (Konikow 2013). The author can only speculate as to other factors that might contribute to the loss of Pine Barrens wetlands—e.g., less precipitation, increasing temperature, plant succession, soil desiccation that in turn increases groundwater infiltra-

---

development and recruitment success; 6) stream fish and macroinvertebrates; 7) ecological processes: nitrogen; 8) ecological processes: indicators of physiological stress; 9) landscape models; 10) build-out and water-demand scenarios; 11) data management and data-analysis coordination; and 12) public information and final Kirkwood-Cohansey assessment (Brown 2004: 3).

4. HUC 11.

5. HUC 14.

tion, withdrawal of water immediately adjacent to a specific local wetland (personal communication, Claude Epstein, October 12, 2018). This noted phenomenon appears to coincide with a proliferation of artesian wells in southern New Jersey (Epstein 1985: 7, 8, Table. 1). Another possibility is increased evapotranspiration caused by reforestation after centuries of overexploitation (Ballard 1979) or longer growing seasons (Wichansky et al. 2006), Large perennial ponds in which residents fished, boated, and swam, dotted the Pinelands a century ago. Little evidence remains that many ever had been ponds. Cripples that in the past mired wagons are today dry hollows. A century ago, strong springs flowed with such great force that they inspired tales of whirlpools, bottomless chasms, and sparkling blue waters. They, too, are gone. So are thousands of acres of savannah wetlands.

Spungs, cripples, blue holes, and savannah may in many instances remain only as recollection, but they are not irrelevant, nor are those that are left expendable. They possess significant potential for future historical and scientific research (Fig. 9.2). The wetlands serve as important habitat for rare and endangered flora and fauna. For certain, we know that the demands placed upon South Jersey's water supplies are increasing. What is unknown is how over-withdrawal from deeper aquifers has affected—and will affect—the ecosystem so dependent upon a ready supply of near-surface water. While this groundwater is but the thinnest of skin atop a tremendous 67-trillion-liters (17.7-trillion-gallons) underground reservoir,[6] it is arguably the most important component of the ecosystem regarding Pine Barrens biodiversity. The shallow aquifers are the lifeblood of the Pinelands. Local hydrologist Claude Epstein (personal communication, February 19, 2021) suggests that these are unconfined aquifers that have a leaky—sometimes very leaky—impervious layer. Such water-bearing formations are often thick and have water recharge them from lower, confined aquifers. So, while their surfaces may dry out seasonally (and perhaps over time), they still draw water from the deeper aquifers.

---

6. A common misquote is that the Kirkwood-Cohansey aquifer contains 17-trillion gallons of pristine groundwater. The correct metric is 17.7-trillion gallons, which is based on analysis by Rhodehamel (1979b), in a refinement by Forman (1979b: 572). Steady Eddy Rhodehamel (1919–2017) passed at the age of 97.

Preservation of Our Periglacial Heritage

The mid-Atlantic region of the USA is experiencing unprecedented development, with rapid extension of low-density housing and commercial functions into previously rural areas. In New Jersey, development pressures are extreme—land developed between 1986 and 1995 in New Jersey consumes double the per-capita acreage of that developed prior to 1986 (Evans 2006). That trend continues today. Valued landscapes go to the highest bidder, and their fate is all too often decided by economic rather than ethical concerns (Harman & Arbogast 2004: 34). Associated problems range from the ecological patchworks created by low-intensity exurban and recreational housing to unsustainable drawdown of groundwater reserves in the New Jersey Pinelands. Problems compound as the Pinelands Commission's planning power decentralizes and shifts to facilitate local entrepreneurial interests (Mason 1992), at times under the guise of Smart Growth.[7] When asked in a survey (Goertzel & Leonardis 2006) what they hated most about South Jersey, residents overwhelmingly cited sprawl.[8] Better recognition, understanding, and appreciation of the many exceptional periglacial sites in the region will become an effective tool in attempts to preserve our collective heritage from exurban sprawl.

Geomorphology is an underutilized tool in land-preservation efforts, particularly in the USA. Geotopes, locations where natural geological or

---

7. In the governance of Pine Barrens development, the Pinelands Commission routinely invokes "Smart Growth" as a guiding principle. Miller and Hoel (2002: 16) reason, "At first glance, who can rightly oppose the notion of 'smart growth', since its opposite would presumably be dumb growth?" Apparently, the term is confounding in its New Jersey sense. Kinsey (2008: 7) wrote: "A 2003 amendment to the Local Development and Redevelopment Law used the term 'smart growth principles' to add an absurdly vague criterion for designation of an 'area in need of redevelopment' that could potentially trigger the exercise of local government's power of eminent domain." Anything within a Pinelands Village or Town can now in theory be redeveloped, including historic structures, wetlands, and habitat with documented threatened and endangered species. In example, Geubtner (2008: 2) provides a list of reasons that land in this author's village was deemed in need of redevelopment: 1) lands have remained vacant and underutilized for a period of ten or more years cannot likely be developed through the instrumentality of solely private capital 2) a lack of roadways servicing the site; 3) soils were too poor for development; 4) wetlands were present; and 5) critical habitat existed.

8. This survey of 444 residents was conducted in the fall of 2000. When asked what they did not like about South Jersey, the interviewees responded: overdevelopment, too many houses, too much traffic—51%; insurance too expensive—27%; taxes too high—26%; not enough to do—19%; pollution—19%; problems of the older cities such as Camden—17%; climate—12%; poor shopping—9%; poor schools—8%; and people not friendly—7%.

geomorphological features worthy of protection exist, were recognized in 1972 at UNESCO's Convention on the Protection of the Cultural and Natural Heritage of the World (Stott 2005). Unlike biotopes, geotopes have not received the attention in the United States that they have in Europe. In Germany alone, 176 national geotopes have been recognized as worthy of protection for tourism and study (Akademie für Geowissenschaften und Geotechnologien 2018) [Update] along with 169 Global geoparks in 44 countries (Nationaler Geopark 2022). Heightened appreciation for the natural history of a region can provide a basis for the preservation of extensive preserves and heritage sites, such as the Ice Age Scientific Reserve in Wisconsin (Black 1974) and the Albany Pine Bush National Natural Monument Pleistocene dune field (Lookingbill et al. 2012).

The formidable cultural landscape of the Pine Barrens is well documented (e.g., Beck 1945; Sinton 1978; Jahn 1980; Moonsammy et al. 1987; Mountford 2002; Demitroff 2014). Two methods of landscape interpretation are available for the investigation of human landscapes, the study of tangible elements of a cultural resource (e.g., houses, barns, fences, land-use systems), or inquiry into people's perception and responses to a cultural resource (Myers et al. 2003). The Pines were settled along a Pinelands ethnic archipelago model (Berger & Sinton 1985: 97–99; Marsh et al. 2019). Within Marsh's study,[9] it is described how the region's unusual geography (i.e., spungs, swamps, relict plateaus [Marsh et al. 2019]) resulted in islands of ethnic settlement. Margaret Mead's childhood association with this ethnic archipelago likely bolstered her adult interest in anthropology (Mead 1927, 1928). In her autobiography, *Blackberry Winter: My Earlier Years*, Mead (1972) recounted how experiences on the family's Pine Barrens farm profoundly influenced her formative years (Chappine & Demitroff 2016).

In their work, Berger and Sinton (1985: xvii–xviii) commented that planners spend much of their time on the physical aspects of a site but fail to relate a location's importance to the cultural ecology. Addressing this problem, they state that cultural geography "can help dispel a planner's arrogance that only he or she really understands a region and how to treat it. The beauty and complexity of the Pine Barrens should awe the officials in charge of the region's future. The deep understanding of any place can leave no room in a planner's mind for cynicism." Perhaps their message should be extended to the scientific community as well.

---

9. Libby Marsh's (1923–2009) son, Ben Marsh, is a Professor of Geography at Bucknell University, Lewisburg, PA, and studies the periglacial environment of Pennsylvania.

Conclusion

It is either distressing or exciting, according to one's nature, to find that the trail one expects to lead toward a certain point may carry one into unexpected directions (Sauer 1956: 288).

This personal and spirited adventure of discovery along the sandy tracks of the Pine Barrens began early in this author's childhood. Experiencing this remarkable landscape on such an intimate level was an important part of my geographical apprenticeship. More than a half-century of field experience helped cultivate insight into the surrounding landforms. Through newfound mature-student training in the intellectual context of geographical dynamics, a story of how southern New Jersey's periglacial legacy influenced land-use patterns for 14,000 years unfolds. Local need for food supplies and later extra-local demands for forest products established a sequential pattern of interaction with periglacial landforms. Ultimately, excessive demands on water resources by an urbanizing population in and (especially) outside the Pinelands are contributing to the apparent degradation of shallow groundwater-dependent wetlands, geographic studies on the complexities of environmental, human/societal, and environmental/societal dynamics abound. I am unaware of any other place where Late Pleistocene periglacial conditions and processes have so profoundly influenced a region's physical, cultural, and biological systems.

Fig. 9.1 Whither Pinelands wetlands past, present, and future?

Pinelands wetlands are ever so slowly fading away. Between 2000 and 2006, over a dozen magazine and newspaper articles chronicled the story of the disappearance of Pinelands wetlands (e.g., clockwise, Hajna 2004; Moore 2002b; Harper 2004; Hajna 2000a; Hajna 2002; Moore 2002a). Are Pinelands wetlands ever so slowly fading away? Reproduced — respectively — with permissions from: ©USA Today Network; ©New Jersey Monthly; ©Press of Atlantic City; ©USA Today Network; ©USA Today Network; ©USA Today Network.

Fig. 9.2 As South Jersey's wetlands fade, history goes with them.

Oblique aerial photograph of an Inner Coastal Plain spung investigated for archeological significance by Bonfiglio and Cresson (1982). Photo taken March 1976 by Anthony Bonfiglio. Johnson Farm, near Vincentown, NJ.

References

Abbott, C. C., 1907: *Archæologia Nova Cæsarea*. Trenton, NJ: MacCrellish & Quigley. 70 pp.

Abbott, C. C., 1918: "The Archæological Significance of an Ancient Dune." *Proceedings of the American Philosophical Society*. **57**, 1: 49–59

Able, K. W., 2020: *Beneath the Surface: Understanding Nature in the Mullica Valley Estuary*. New Brunswick, NJ: Rutgers University Press. 309 pp.

Ackley, J. A., 1945: "Notes on the Black Water Stream and Malaga Road." *The Vineland Historical Magazine*. **30**, June-April: 159–161.

Adlum, J., and Wallis, J., 1791: *Map exhibiting a general view of the roads and inland navigation of Pennsylvania, and part of the adjacent states: Respectfully inscribed to Thomas Mifflin, governor, and the General Assembly of the commonwealth of Pennsylvania*. John Adlum, and John Wallis, Philadelphia: s.n. Scale 1:650,000.

Akademie für Geowissenschaften und Geotechnologien, 2018: Herzlich Willkommen. https://www.geoakademie.de/index.php Accessed August 08, 2018.

Åkerman, J., 1982: "Studies on naledi (icings) in West Spitsbergen." In French, H. M. (ed.), *The Roger J. E. Brown Memorial Volume. Proceedings of the Fourth Canadian Permafrost Conference, Calgary, Alberta, March 2–6, 1981*. Ottawa: National Research Council of Canada, pp. 189–202.

Alekseyev, V. R., 1998: "Water-Heat Budget and Ecological Structure of Icing Landscape Complexes." *Polar Geography*. **22**, 3: 211–221.

Alekseyev, V. R., 2015: "Cryogenesis and geodynamics of icing valleys." *Geodynamics & Tectonophysics*. **6**, 2: 171–224. (In English).

Alexander, R. C., 1951: "Bennett Bog." *The Cape May Geographic Society Annual Bulletin*. **5**: 10–16.

Allcroft, A. H., 1908: *Earthwork of England: Prehistoric Roman, Saxon, Danish, Norman & Mediæval*. London: Macmillan. 711 pp.

Allcroft, A. H., 1924: *Downland Pathways*, 2nd edition. London: Mathuen. 292 pp.

Ambert P., Clauzon G. 1992. "Morphogens éolienne en ambiance périglaciaire: le dépressions fermées du pourtour du Golfe du Lion (France méditerranéenne)." *Zeitschrift für Geomorphologie*, Neue Folge, Supplementband. **84**: 55–71.

Anderson, H. A., 1964: "Farmington." *In* Egg Harbor Township Tercentenary Publications Committee, *Sketches of Old Egg Harbor Township*. Egg Harbor City, NJ: Laureate Press. pp. 66–69.

André, Marie-Francoise, 2003: "Do Periglacial Landscapes Evolve under Periglacial Conditions?" *Geomorphology*. **52**: 149–164.

Anketell, J. M., Cegla, J., and Dzulynski, S., 1970: "On the Deformational Structures in Systems with Reversed Density Gradients." *Rocznik Polskiego Towarzystwa Geologicznego/ Annales de La Société Géologique de Pologne*. **40**, 1: 3–30.

Anonymous, 1879. "May's Landing Glass-Sand Pits." *Mays Landing Record*. **2**, 21: 2, Saturday, March 08.

Anonymous, 1880: "Tar Kiln Neck." *Mays Landing Record*. **3**, 37: 3, Saturday, June 26.

Anonymous, 1893: "The Truth Because It's Sworn to a Night of Horror in the New Jersey Pine Barrens." *Philadelphia Inquirer*. **128**, 120, April 30, 1893: 21.

Anonymous, 1908a: "Captain Kidd Treasure." *The Evening Journal*. August 14 (reprinted 1982 in *The Vineland Historical Magazine*. **58**, 1: 3).

Anonymous, 1908b: "Curious Find." *The Evening Journal*. August 14 (reprinted 1982 in *The Vineland Historical Magazine*. **58**, 1: 47).

Arsenault, J. R., in press: "The Shape and Age of a Southern New Jersey Spung." Bartonia, Journal of the Philadelphia Botanical Club. **72**: XX–XX.

Baer, C., 1944: "Life in Southern New Jersey Furnace Community in Early 19th century." *Steel Facts*. **64**, February–March: 6–7.

Bailey, S., (ed.), 1999: "Aged Millville Man Recalls Death of Last of South Jersey Bears." Reprint from *The Bridgeton Evening News*, September 25, 1929. *South Jersey Magazine*. **28**, 4, Winter Issue: 12.

Bakker, J. P., Esselink, P., Dijkema, K. S., van Duin, W. E., and de Jong, D. J., 2002: "Restoration of Salt Marshes in the Netherlands." *Hydrobiologia*. **478**: 29–51.

Ballantyne, C. K., 2018: *Periglacial Geomorphology*. Hoboken, NJ: Wiley Blackwell. 472 pp.

Ballard, J. T., 1979: "Fluxes of Water and Energy hrough the Pine Barrens Ecosystems." *In* Forman, R. T. T. (ed.), *Pine Barrens: Ecosystem and Landscape*. New York: Academic Press. pp. 133–146.

Barber, J. W., and Howe, H., 1868: *Historical Collections of New Jersey: Past and Present*. Newark. 544 pp. (1975 reprint, The Reprint Company, Spartanburg, SC).

Barksdale, H. C., 1952: "Ground Water in the New Jersey Pine Barrens Area." *Bartonia, Journal of the Philadelphia Botanical Club*. **26**: 36–38.

Barrale, M., 1979: *History of the Borough of Buena*. Unpublished manuscript. 114 pp. (reprinted 2021 through The Create Space, 158 pp.)

Barrows, H. H., 1923: "Geography as Human Ecology." *Annals of the Association of American Geographers*. **13**, 1: 1–14.

Bassett, T. J., and Zimmerer, K. S., 2003: "Cultural Ecology." *In* Gaile, G. L., and Willmott, C. J. (eds.). *Geography in America at the Dawn of the 21st Century*. Oxford: Oxford University Press. pp. 97–112.

Bauman, E., and Cassaday, H., 1999: *The History of Methodism for Friendship United Methodist Church*. Landisville, NJ: Friendship United Methodist Church. 21 pp.

Beans, B. E., and Niles, L. (eds.), 2003: *Endangered and Threatened Wildlife of New Jersey*. New Brunswick, NJ: Rutgers University Press. 328 pp.

Beck, H. C., 1937: *More Forgotten Towns of Southern New Jersey*. New

Brunswick, NJ: Rutgers University Press. 338 pp.

Beck, H. C., 1945: *Jersey Genesis: The Story of the Mullica River.* Rutgers University Press: New Brunswick, NJ: Rutgers University Press. 304 pp.

Becker, D. W., 1964: *Indian Place Names in New Jersey.* Cedar Grove, NJ: Phillips-Campbell Publishing. 111 pp.

Beckett, A., 1909: *The Spirit of the Downs: Impressions and Reminiscences of the Sussex Downs.* London: Methuen. 368 pp.

Beers, F. W., 1872a: *State Atlas of New Jersey, Based on State Geological Survey and From Additional Surveys.* New York: Beers, Comstock & Cline. 122 pp.

Beers, F. W., 1872b: *Topographic Map of Atlantic County, New Jersey, from recent and actual surveys.* New York: Beers, Comstock & Cline. 1 sheet.

Beers, S. N., Beers, F. W., Lake, L. B., and Warner, C. S., 1862: *Map of Cumberland County, New Jersey.* Philadelphia. PA: A. Pomeroy. 1 sheet.

Bélisle, A. C., Asselin, H., LeBlanc, P., and Gauthier, S., 2018: "Local Knowledge in Ecological Modeling." *Ecology and Society.* **23**: 1–11.

Bennett, E., 1999: "On Joy, Sorrow and the Blue Holes of Area Pinelands." *The Press of Atlantic City.* October 31, **XCIX**, 304: A1, A15.

Berger, J., and Sinton, J. W., 1985: *Water, Earth, and Fire: Land Use and Environmental Planning in the New Jersey Pine Barrens.* Baltimore, MD: The John Hopkins University Press. 228 pp.

Berón, M.A., 2016: "Dunes, Hills, Waterholes, and Saltpeter Beds: Attractors for Human Populations in Western Pampa, Argentina." *Quaternary International.* **422**: 163–173.

Bien, W. F., 1999: *Ecological Factors Influencing Spatial Patterns in Sphagnum flavicomans and Sphagnum pulchrum from the New Jersey Pine Barrens.* Ph.D. dissertation, Philadelphia, PA: Drexel University. 199 pp.

Bisbee, H. H., 1971: *Sign Posts in History of Burlington County, New Jersey.* Willingboro, NJ: Alexia Press. 304 pp.

Bishop, B., 1831: Resurvey and map of 164.11 acres near Towns Pond, Galloway Township, Gloucester County.

Bizub, R., 2008: "Growth Areas: a Water Quality Conundrum." *Inside the Pinelands.* **15**, 2, Special Water Issue: 4.

Bizub, R., 2010: "Water Supply Challenges—Cape May County!" *Pinelands Watch.* **53**: 1–3.

Bizub, R., 2014a: "Wells Can Suck the Pinelands Dry." *Inside the Pinelands.* **21**, 4: 4.

Bizub, R., 2014b: "Groundwater—An Environmental Issue." *Inside the Pinelands* **21**, 4, June/July: 1, 5.

Black, R. F., 1974: *Geology of Ice Age National Scientific Reserve of Wisconsin.* Scientific Monograph No. 2, National Park Service, 234 pp.

Black, R. F., 1983: "Pseudo-Ice-Wedge-Casts of Connecticut, Northeastern United States." *Quaternary Research.* **20**: 74–89.

Black, R. F., and Barksdale, W .L., 1949: "Oriented Lakes of Northern Alaska." *Journal of Geology.* **57**: 105–118.

Blackman, L., 1880: *History of Little Egg Harbor Township, Burlington County, NJ.* Tuckerton, NJ: The Great John Mathis Foundation. 291 pp. (1963 reissue).

Blinn, M. E., 1918: "The Old Time Tar Kilns." *Vineland Historical Magazine.* **3**: 29–30.

Boardman, J., (ed.), 1987: *Periglacial Processes and Landforms in Britain and Ireland.* Cambridge: Cambridge University Press. 296 pp.

Bockheim, J. G., and Hartemink, A. E., 2013: "Soils with Fragipans in the USA."

Catena. **104**: 233–242.

Bockheim J. G. 2014. *Soil Geography of the USA: A Diagnostic-Horizon Approach*. Heidelberg, DE: Springer Cham. 319 pp.

Bonfiglio, A., 1985: "Amerindian Utilization of Barchan, Parabolic and Longitudinal Dunes in Southern New Jersey (A Preliminary Investigation)." Paper presented at the Fourth Annual Pinelands Symposium, Easthampton, N.J.

Bonfiglio, A., and Cresson, J.A. 1978: "Aboriginal Cultural Adaptation and Exploitation of Periglacial Features in Southern N.J." paper presented at Eastern States Archaeological Federation meeting, Bellmawr, N.J.

Bonfiglio, A., and Cresson, J.A., 1982: "Geomorphology and Pinelands Prehistory: a Model into Early Aboriginal Land Use." In Sinton, J. W. (ed.), *History, Culture and Archeology of the New Jersey Pine Barrens*. Center for Environmental Research, Pomona, NJ: Stockton State College. pp. 18–67.

Boonman, J. G., and Mikhalev, S. S., 2005: "The Russian Steppe." In Suttie, J. M., Reynolds, S. G., and Batello, C., *Grasslands of the World*. Rome: Food and Agriculture Organization of the United Nations. 514 pp.

Boreham S., 1996. "Possible Termokarst Landforms near Theydon Bois, Essex." *London Naturalist*. **75**: 21–26.

Boreham, S., 1997: "Holocene (Flandrian) Organic Deposits Preserved in an Icing Hollow at Bardfield Bridge, Borley, Essex." *Quaternary Newsletter*. **82**: 9–14.

Boreham, S., and Horne, D. C., 1999: "The Role of Thermokarst and Solution in the Formation of Quidenham Mere, Norfolk, Compared with Some other Breckland Meres." *Quaternary Newsletter*. **87**: 16–27.

Boreham, S., White, T. S., Bridgland, D. R., Howard, A. J., and White, M. J., 2010: "The Quaternary History of the Wash Fluvial Network, UK." *Proceedings of the Geologists' Association*. **121**: 393–409.

Borisova, O., Sidorchuk, A., and Panain, A. 2006: "Palaeohydrology of the Seim River Basin, Mid-Russian Upland, Based on Palaeochannel Morphology and Palynological Data." *Catena*. **66**: 53–73.

Boucher, J. E., 1963: *Absegami Yesteryear*. Egg Harbor City, NJ: Laureate Press. 149 pp.

Bowen, F. W., 1885: *History of Port Elizabeth, Cumberland County, New Jersey, Down to the Present Time*. Millville, NJ: Millville Publishing. 104 pp. (reprinted and expanded 1936).

Bowman, J. F., 1966: *Petrology of the Pensauken Formation*. Ph.D. dissertation, New Brunswick, NJ: Rutgers University, 155 pp.

Bowman, J. F., and Lodding, W., 1969: "The Pensauken Formation—a Pleistocene Fluvial Deposit in New Jersey." In Subitzky, S. (ed.), *Geology of Selected Areas in New Jersey and Eastern Pennsylvania and Guidebook of Excursions*. New Brunswick, NJ: Rutgers University Press. pp. 3–6.

Boyd, H. P., 1991: *A Field Guide to the Pine Barrens of New Jersey*. Medford, NJ: Plexus Press. 423 pp.

Boyd, H. P., 2008: *The Ecological Pine Barrens of New Jersey: An Ecosystem Threatened by Fragmentation*. Medford, NJ: Plexus Publishing. 272 pp.

Boyer, C. S., 1931: *Early Forges & Furnaces in New Jersey*. Philadelphia: University of Pennsylvania Press. 287 pp.

Boyer, C. S., 1962: *Old Inns and Taverns in West Jersey*. Camden, NJ: Camden County Historical Society. 326 pp.

Braddock-Rogers, K., 1930: The bog ore industry in South Jersey prior to 1845. *Journal of Chemical Education*. **7**, 7:

1493-1519.

Bridges, N. T., and Laity, J. E., 2013: "Fundamentals of Aeolian Sediment Transport: Aeolian Abrasion. Reference Module in Earth Systems and Environmental Sciences." *Treatise on Geomorphology*. **11**, 2013: 134-148.

Brissot de Warville, Jacques-Pierre, 1792. *New Travels in the United States of America*. London: Printed for J. S. Jordan. 483 pp.

Brooks, R. T., 2004: "Weather-Related Effects on Woodland Vernal Pool Hydrology and Hydroperiod." *Wetlands*. **24**, 1: 104-114.

Brown, A. H., 1879: "The Character and Employments of the Early Settlers of the Sea-Coast of New Jersey." Prepared at the Request of the New Jersey Historical Society, and Read at their Meeting in the City of Newark, May 15th, 1879. Newark, NJ: Newark Daily Advertiser. 37 pp.

Brown, A. M., 2004: "Aquifer Study in Full Swing." *New Jersey Flows*. New Jersey Water Resources Research Institute Special Pinelands Edition. **5**, 2: 2-3.

Brown, R. H., 1943: *Mirror for Americans: Likeness of the Eastern Seaboard, 1810*. New York: American Geographical Society. 312 pp.

Brown, R. J. E., and Péwé, T. L., 1973: "Distribution of Permafrost in North America and Its Relationship to the Environment: a Review, 1963-1973." *In* National Research Council of Canada. *Permafrost: The North American Contribution to the Second International Conference, Yakutsk*. Washington, DC: National Academy of Sciences. pp. 71-100

Brunnschweiler, D., 1962: "The Periglacial Realm in North America During the Wisconsin Glaciation." *Biuletyn Peryglacjalny*. **11**: 15-27

Büdel, J., 1944: „Die Morphologischen Wirkungen des Eiszeit-klimas im Gletscherfreien Gebiet." *Geologische Rundschau*. **34**: 482-519.

Büdel, J., 1982: *Climatic Geomorphology*. translated by Lenore Fischer and Detlef Busche, Princeton, NJ: Princeton University Press. 443 pp.

Bull, A. J., 1939: "Nivation in the South Downs: part II, the Dry Valleys." *Association for the Study of Snow and Ice*. **1**, 3: 21-22.

Bunnell, J. F., Laidig, K. J., Burritt, P. M., and Sobel, M. C., 2018: *Vulnerability and Comparability of Natural and Created Wetlands*. Final report to the U.S. Environmental Protection Agency, Pinelands Commission, New Lisbon, New Jersey, USA. 58 pp.

Burke, R. T. A., Thorp, J., and Selzer, W. G., 1929: "Soil Survey of Salem Area, New Jersey." Series 1923, Number 47. Bureau of Chemistry and Soils. Washington, DC: United States Department of Agriculture. pp. 1649-1696.

Burr, D. H., and Illman, T., 1835: "New-Jersey." Entered . . . Burr, David H., *A New Universal Atlas; Comprising Separate Maps of All the Principal Empires, Kingdoms & States Throughout the World: and forming a distinct Atlas of the United States*. New York: D. S. Stone. Scale 1: 918,720.

Burrough, J., Champion, J., Sayers, D., Wescott, R., Lucas, S., and Johnson, W., 1793: A road from the Blue Anchor to the sea shore. Recorded 28th October, 1793. 3 pp.

Burton, I., 1963: "The Quantitative Revolution and Theoretical Geography." *Canadian Geographer*. **7**, 4: 151-162.

Cailleux, A., 1936: "Les Actions Éoliennes Périglaciaires en Europe (Quaternary Periglacial Wind Action in Europe)." *Bulletin de la Société Géologique de France*. **5**: 495-505.

Cailleux, A., 1961: "Mares et Lacs Ronds et Loupes de Glace du Sol." *Biuletyn Peryglacjalny*. **10**: 35-41.

Cain, O. B., 1958: *The History of Mays Landing and Vicinity*. Graduate term paper, Glassboro State College, Glassboro, NJ. 43 pp.

Calhoun, A. J. K., deMaynadier, P. G. (eds.), 2008: *Science and Conservation of Vernal Pools in Northeastern North America*. Boca Raton, FL: CRC Press. 363 pp.

Calhoun, A. K. J., Mushet, D. M., Bell, K. P., Boix, D., Fitzsimons, J. A., and Isselin-Nondedeug, F., 2017: "Temporary Wetlands: Challenges and Solutions to Conserving a 'Disappearing' Ecosystem." *Biological Conservation*. **211**, Part B, July 2017: 3–11.

Carlson, A. E., Dutton, A., Long, A. J., and Milne, A. J., 2019: "PALeo Constraints on SEA Level Rise (PALSEA): Ice-Sheet and Sea-Level Responses to Past Climate Warming." *Quaternary Science Reviews*. **212**: 29–32.

Carozza, J.-M., Llubes, M., Danu, M., Faure, E., Carozza, L., David, M., and Manen, C., 2016: "Geomorphological Evolution of Mediterranean Enclosed Depressions in the Late Glacial and Holocene: The Example of Canohès (Roussillon, SE France)." *Geomorphology*. 273: 78–92.

Carson, C. E., and Hussey, K. M., 1962: "The Oriented Lakes of Arctic Alaska." *Journal of Geology*. **70**, 4: 417–439.

Castañeda, C., Javier, G., Rodríguez-Ochoac, R., Zarroca, M., Roque, C., Linares, R., and Desir, G., 2017: "Origin and Evolution of Sariñena Lake (Central Ebro Basin): A Piping-Based Model." *Geomorphology*. **290**: 164–183.

Castel, I. I. Y., 1991: "Late Holocene Eolian Drift Sands in Drenthe (The Netherlands)." *Nederlandse Geografische Studies*. **133**: 1–156

Catt, J. A., 1987: "Effects of the Devensian Cold Stage on Soil Characteristics and Distribution in Eastern England." In Boardman, J. (ed.), *Periglacial Processes and Landforms in Britain and Ireland*. Cambridge: Cambridge Univ. Press. pp. 145–152.

Catt, J. A., Green, M., and Arnold, N. J., 1982: "Naleds in a Wessex Downland Valley." *Proceedings of the Dorset Natural History and Archaeological Society*. **102**: 69–75.

CFWNJ (Conserve Wildlife Foundation of New Jersey), 2021: *New Jersey Endangered and Threatened Species Field Guide*. Search our Field Guide for in-depth information about New Jersey's endangered, threatened, and special concern species. http://www.conservewildlifenj.org/species/fieldguide/ Accessed February 14, 2021.

Chalmers, K. H., 1951: *Down the Long-a-Coming*. Moorestown, NJ: The News Chronicle. 206 pp.

Chamberlain, G. A., 1925; "Jarred, Last of the Pineys." *Saturday Evening Post*. **197**, 48, May 30: 113–121 (Reprinted in *SoJourn: A Journal Devoted to the History, Culture, and Geography of South Jersey*. **6**, 1: 51–59.

Chamberlain, T. C., 1897: Editorial footnote. *Journal of Geology*. **5**: 826.

Chappine, P., and Demitroff, M., 2016: "Where Blackberries Grew: Margaret Mead in Hammonton," *SoJourn: A Journal Devoted to the History, Culture, and Geography of South Jersey*. **1**, 2: 37–43.

Chew, R., 1815: Deed to 48.5 acres at a place called the Park. West Jersey Loose Records (49956).

Chmielewski, R. M., 1996: *Hydrologic Analysis of Carolina Bay Wetlands at the Savannah River Site, South Carolina*. M.S. thesis, University of Wisconsin, Milwaukee, WI.

Christiansen, H. H., Matsuoka, N., and Watanabe, T., 2016: "Progress in Understanding the Dynamics, Internal Structure and Palaeoenvironmental Potential of Ice Wedges and Sand

Wedges." *Permafrost and Periglacial Processes.* **27**, 4: 365–376.

Clark, B., and Hitchman, G., 1802: Resurvey and map of 6,156 acres of the Lake Tract, Gloucester County. West Jersey Loose Records, 1802, 53576.

Clement, J., 1845: *A Collection of Maps and Drafts Copied from Old Maps & Drafts and from Actual Surveying by John Clement of Haddonfield, December 17th 1845.* Volume First, Volume Two in back. Clement Papers, Maps & Draughts ,Vol. 2 (Gen C1 3.1), The Historical Society of Pennsylvania, Philadelphia, PA. 104 pp. + index (2 pp.).

Clement, J., 1852: *A Collection of Maps and Drafts Copied from Old Maps & Drafts and from Actual Surveying by John Clement, October 16, 1852.* Volume Third, Volume Four in back. Clement Papers, Maps & Draughts, Vols. 3&4 (Gen C1 3.2), The Historical Society of Pennsylvania, Philadelphia, PA. 90 pp. + index (3 pp.).

Clement, J., 1867: *A Collection of Maps and Draughts Copied from Old Maps & Draughts and from Actual Surveying by John Clement of Haddonfield, August 1867.* Volume Sixth Clement Papers, Maps & Draughts (Gen C1 3.1), The Historical Society of Pennsylvania, Philadelphia, PA. 88 pp. + index (3 pp.).

Clement, J., 1880: Atlantic County, "In Ye Olden Time"—and Now. *In* Surveyors' Association of West New Jersey. *Proceedings, Constitution, By-Laws, List of Members, &c., of the Surveyors' Association of West New Jersey: With Historical and Biographical Sketches Relating to New Jersey.* Camden, NJ: published by order of the Society. pp. 401–420.

Clement, J., 1888: "Charles Brockden." *Pennsylvania Magazine of History and Biography.* **12**: 185–189.

Cohen, D. S., 1983: *The Folklore and Folklife of New Jersey.* New Brunswick, NJ: Rutgers University Press. 253 pp.

Cohen-Zada A. L., Blumberg D. G., Maman S. 2016. Review Article: "Earth and Planetary Aeolian Streaks: A review." *Aeolian Research.* **20**: 108–125.

Collin, G., and Godard, A., 1962: "Les Dépressions Fermées en Lorraine" (The closed depressions in Lorraine). *Revue Géographique de 'Est.* **3**: 233–261.

Collins, B. R. and Anderson K. H., 1994: *Plant Communities of New Jersey: A Study in Landscape Diversity.* New Brunswick, NJ: Rutgers University Press. 287 pp.

Colonial Williamsburg Foundation, 2018: *A Colonial Gentlemen's Clothing: a Glossary of Terms.* http://www.history.org/history/clothing/men/mglossary.cfm. Accessed August 10, 2018.

Conegno, C., 2014: Report: "Pinelands Water Supply Takes Big Hit." *Courier Post.* August 8.

Cook, G., 1966: "Notes on South Jersey Railroad Names, Gauges, Charters, Opening dates, Consolidations, and Mergers or Purchases." Adapted for the Buena Historical Society by Mark Demitroff, July 5, 1999, 8 pages.

Cook, G. H., 1857: *Geology of the County of Cape May, State of New Jersey.* Trenton, NJ: Office of the True American. 211 pp.

Cook, G. H., and Vermeule, C. C., 1890: *Atlas of New Jersey.* NJ Geological Survey, New York: Julius Bien. 20 sheets.

Cook College Office of Continuing Professional Education Programs, 2006: Wetlands training spring 2006: basic wetlands training, wetlands delineation certificate series, advanced wetlands training. New Brunswick, NJ: The State University of New Jersey Rutgers. 16 pp.

Cooper, R. A., 2002: "Scientific Knowledge of the Past is Possible: Confronting Myths about Evolution

& Scientific Methods." *The American Biology Teacher.* **64**, 6: 427–432.

Cope E. D. 1896. "Fishes in Isolated Pools." *The American Naturalist.* **359**: 943–944.

Côte, M. M., and Burn, C. R., 2002: "The Oriented Lakes of Tuktoyaktuk Peninsula, Western Arctic Coast, Canada: a GIS-Based Analysis." *Permafrost and Periglacial Processes.* **13**: 61–70.

Cottrell, A. T., 1937: "Unused South Jersey—Its Vast Latent Possibilities. What Can Be Done with 1875 Square Miles of Tree-Covered Wilderness, with Magnificent State-Built Roads and Great Railroads Crossing It." *Journal of Industry & Finance, Newark, N.J.* August 1937: 10.

Coxon, P., 1978: "The First Record of a Fossil Naled in Britain." *Quaternary Newsletter.* **24**: 9–11.

Craig, P. S., and Williams, K-E., 2007: *Colonial records of the Swedish Churches in Pennsylvania, Vol. 3—The Sandel years, 1702–1719.* Philadelphia, Pa: Swedish Colonial Society, p. 68.

Cresson, J. A., 1983: "Geomorphology and Pinelands Prehistory; Lithic Resources, Landforms and Human Settlements." Paper presented at the Second Annual Pinelands Symposium, Stockton State College, N.J.

Cresson, J. A., 2000: "Periglacial Landforms on the Coastal Plain." *Newsletter, The Archeological Society of New Jersey.* **191**: 2–4.

Cresson, J. A., and Mounier, R.A., 2009: "Prehistoric Periglacial Landform Use Redo: an Example of Archaeological Predictability and Confirmation of Deeply Buried Prehistoric Remains within Aeolian Sands at a Site in Evesham Township, Burlington County." *Newsletter of the Archaeological Society of New Jersey.* **226**: 1–7.

Cresson, J. A., Mounier, (R.) A., Bonfiglio, A., and Demitroff, M., 2006: "Periglacial Landforms of Southern New Jersey: Sites, Trails and Ancient Cultural Links." *In* Hellström, R., Frankenstein, S. (eds.), *Program and Abstracts, 63rd Eastern Snow Conference, University of Delaware, Held Jointly with the Cryosphere Specialty Group of the Association of American Geographers, 7–9 June 2006,* p. 72.

Cresswell, T., 2014: "Déjà Vu All Over Again: Spatial Science, Quantitative Revolutions and the Culture of Numbers." *Dialogues in Human Geography.* **4**, 1: 54–58.

Cromartie, W. J. (ed.), 1982: *New Jersey's Threatened and Endangered Plants and Animals.* Pomona, NJ: Richard Stockton College. 385 pp.

Cunnington, C. W., 1972: *Handbook of English Costume in the Eighteenth Century.* Boston: Plays. 453 pp.

Cushing, T., and Sheppard, C. E., 1883: *History of the Counties of Gloucester, Salem, and Cumberland, New Jersey with Biographical Sketches of Their Prominent Citizens.* Philadelphia, PA: Everts & Peck. 728 pp.

Custer, J. F., 1986: "Periglacial Features of the New Jersey Coastal Plain: a Demurrer." *The Bulletin of the Archaeological Society of New Jersey.* **40**: 21–25.

Czudek, T., 1986: "Some Problems of Valley Development in Pleistocene Deposits of Northern Moravia (Czechoslovakia)." *Biuletyn Peryglacjalny.* **31**: 35–45.

Dalgaard, Carl-Johan, Kaarsen, N., Olsson, O., and Selaya, P., 2018. "Roman Roads: A Persistent Effect of Infrastructure on Development." VOX CEPR Policy Portal. 10 April 2018. https://voxeu.org/article/roman-roads-and-persistence-development Accessed August 07, 2018.

Das, S., 2007: "Origin and Evolution of

Dry Valleys South of Ronkonkoma Moraine, Long Island." M.S. thesis, Stony Brook University, Stony Brook, NY, 57 pp.

Day, M., Anderson, W., Kocurek, G., and Mohrig, D., 2016: "Carving Intracrater Layered Deposits with Wind on Mars." *Geophysical Research Letters*. **43**, doi:10.1002/2016GL068011.

de Blij, H., 2005: *Why Geography Matters. Three Challenges Facing America: Climate Change, The Rise of China, and Global Terrorism.* Oxford: Oxford University Press. 308 pp.

Del Sontro, T. S., 2003: "Spungs of the Great Egg Harbor Watershed: Mapping Periglacial Features Using Geographical Information System (GIS) and Global Positioning System (GPS)." Methods [B.S. Marine Science Program Distinction Report] Pomona, NJ: Richard Stockton College. 18 pp. plus addenda (project supervised by Mark J. Mihalasky).

Del Tredici, P., 2010: *Wild Urban Plants of the Northeast: a Field Guide.* Ithaca & London: Comstock Publishing Associates. 374 pp.

Delaware, 1845: *Laws of the State of Delaware Passed at a Session of the General Assembly, Commenced and Held at Dover, on Tuesday the Seventh Day of January, in the Year of our Lord, One Thousand Eight Hundred and Forty-Five, and of the Independence of the United States, The Sixty-Ninth.* Dover, DE: S. Kimmel. 125 pp.

Delaware, 1903. *Original Land Titles in Delaware Commonly Known as the Duke of York Record: Being an Authorized Transcript From the Official Archives of the State of Commissions, Surveys, Plats and Confirmations by the Duke of York and Other High Officials, from 1646 to 1679.* Wilmington, Del.: Sunday Star Print, 112 pp.

DELDOT (Delaware Department of Transportation), 1993: Archaeology/Historic Preservation. Archaeological Discoveries in Connection with Scarborough Road Data Recovery Excavations at the Blueberry Hill Prehistoric Site (7K-C-107) Dover, Kent County, Delaware, DelDOT Archaeology Series No. 130, https://www.deldot.gov/archaeology/scarborough_rd/index.shtml?dc=toc Accessed August 12, 2018.

Dellomo, Jr., A. N., 1977: *Harrisville: A Journey Down the Sugar Sand Roads of Yesteryear.* Atlantic City, NJ: Angelo Publishing. 150 pp.

DeLyser, D., and Starr, P. F. (eds.), 2001: "Doing fieldwork." *Geographical Review, Special Issue.* **91**, 1 & 2: 1–508.

DeLyser, D., and Sui, D., 2014: "Crossing the Qualitative–Quantitative Chasm III: Enduring Methods, Open Geography, Participatory Research, and the Fourth Paradigm." *Progress in Human Geography.* **38**, 2: 294–307.

Demitroff, M., 2003: "A Geography of Spungs and Some Attendant Hydrological Phenomena on the New Jersey Outer Coastal Plain." In Hozik, M. J., Mihalasky, M. J. (eds.), *Field Guide and Proceedings, 20th Annual Meeting of the Geological Association of New Jersey, October 10-11, 2003.* Trenton, NJ: Geological Association of New Jersey. pp. 51–78.

Demitroff, M., 2007. "Expanding Our Appreciation and Understanding of Pinelands Natural History." *Pinelands Watch.* **SE-11** (Special Edition), May 2007: 1–3. (*slightly modified and reprinted as* "The Newtonville Dune Field: Expanding Our Appreciation and Understanding of Pinelands Natural History." *New Jersey Audubon.* **Fall/Winter 2007–2008** [Special Places Issue]: 32–33).

Demitroff, M., 2010: "Lakehurst Soil Study. Preliminary Review of Magnetic Lineations – C-17 Assault-

Landing Runway, Joint Air Base McGuire-Dix-Lakehurst, NJ: Potential Environmental Significance." Internal Report, Buckhorn Garden Service, Inc. (Vineland), NJ, 61 pp.

Demitroff, M., 2014: "Sugar Sand Opportunity: Landscape and People of the Pine Barrens." Feature Article, *Vernacular Architecture Newsletter*, Summer 2014. Reproduced in the Vernacular Architecture Newsletter courtesy of NJ VAF 2014. http://vafnewsletter.blogspot.com/2014/07/sugar-sand-opportunity-landscape-and.html. Accessed August 10, 2018.

Demitroff, M., 2016: "Pleistocene Ventifacts and Ice-Marginal Conditions, New Jersey, USA." *Permafrost and Periglacial Processes*. Paleoenvironment Special Issue. **27**: 123–137.

Demitroff, M., Cicali, M., Smith, J., and Demitroff, A. N., 2012: "Ancient Eolian Landforms and Features from a Terrestrial Mid-Latitude Periglacial Environment." In Titus, T. (convener), *Third International Planetary Dunes Workshop: Remote Sensing and Data Analysis of Planetary Dunes*. LPI Contribution No. 1673, Lunar and Planetary Institute, Houston, TX. pp. 29–30.

Demitroff, M., Doolittle, J. A., and Nelson, F. E., 2008: "Use of Ground-Penetrating Radar to Characterize Cryogenic Macrostructures in Southern New Jersey, USA." In Kane, D. L., Hinkel, K. M. (eds.), *Proceedings: Ninth International Conference on Permafrost* (2 Vols.). Institute of Northern Engineering, Fairbanks, AK: University of Alaska Fairbanks. pp. 355–360.

Demitroff, M., Rogov, V. V., French, H. M., Konishchev, V. N., Streletskiy, D. A., Lebedeva-Verba, M. D., and Alekseeva, V. A., 2007: "Possible Evidence for Episodes of Late-Pleistocene Cryogenic Weathering, Southern New Jersey, Eastern USA" (extended abstract). *In* Russian Academy of Sciences, *Proceedings, Vol. II: Cryogenic Resources of Polar Regions, Salekhard City, Polar Cycle, West Siberia, June 2007*. pp. 139–141.

Demitroff, M., Wolfe, S. A., Woronko, B., Chemieloska, D., and Cicali, M., 2019: "Late Pleistocene Ice-Marginal Dune Fields on the Atlantic Coastal Plain, New Jersey Pine Barrens, USA." T4. Eolian Processes and Landscape Evolution (Posters), Geological Society of America Abstracts with Programs. Vol. 51, No. 5, doi: 10.1130/abs/2019AM-336443

Denny, T., 1774: A draft of the New Jersey Society lands in the County of Gloucester on both sides of the Great Egg Harbor River by Thos. Denny, Dept. Surveyor. Survey made 5 May 1774. Lib. T, p. 45. 1 sheet.

Denny, C. S., and Owens, J. P., 1979: "Sand Dunes on the Central Delmarva Peninsula, Maryland and Delaware." Professional Paper 1067-C, US Geological Survey. 15 pp.

Dijkmans, J. W. A., and Koster, E. A., 1990: "Morphological Development of Dunes in a Subarctic Environment, Central Kobuk Valley, Northwestern Alaska." *Geografiska Annaler*. **72A**: 93–109.

Dillingham, (Mr.), 1911: *Immigrants in Industries (in Twenty-Five Parts), Part 24: Recent Immigrants in Agriculture (in Two Volumes: Vol. I)*. Document No. 633, Reports of the Immigration Commission, Washington, DC. 580 pp.

Dinwiddie, C. L., McGikkis, R. N., Stillman, D. E., and Bjella, K. L., 2014: "Interpretation of Geophysical Interrogation of the Great Kobuk Sand Dunes, Arctic Alaska: A Terrestrial Analog Study." Eighth International Conference on Mars, held July 14–18, 2014 in Pasadena, California. LPI Contribution No. 1791, p.1256.

Domber, S., Pallis, T., and Bousenberry, R., 2016: "Springs of New Jersey: Characterization and Assessment." NJ Water Monitoring Council Meeting. Technical Session Theme—Water Monitoring. New Jersey Department of Environmental Protection. http://www.state.nj.us/dep/wms//Domber%20-%20NJ%20Springs%20May%202016.pdf Accessed April 03, 2018.

Dorn, R. I., Krinsley, D. H., Langworthy, K. A., Ditto, J., and Thompson, T. J., 2013: "The Influence of Mineral Detritus on Rock Varnish Formation." *Aeolian Research.* **10**: 61–76.

Douglas, T. A., Fortier, D., Shur, Y. L., Kanevskiy, M. Z., Guo, L., Cai, Y., and Bray, M. T., 2011: "Biogeochemical and Geocryological Characteristics of Wedge and Thermokarst-Cave Ice in the CRREL Permafrost Tunnel, Alaska." *Permafrost and Periglacial Processes.* **22**: 120–128.

Dovey, R. S., 1977: "Groundwater Resources and Geology of Atlantic County: a Primer." Groundwater Subplan Report No. 1, Atlantic City, NJ: Atlantic County 208 Areawide Water Quality Management Planning Agency. 33 pp.

Dowhan, J., Halavik, T., Milliken, A., MacLachlan, A., Caplis, M., Lima, K., and Zimba, A., 1997: "Significant Habitats and Habitat Complexes of the New York Bight Watershed." New Jersey Pinelands COMPLEX #2. U.S. Fish and Wildlife Service, Southern New England-New York Bight Coastal Ecosystems Program, Charlestown, Rhode Island. www.nrc.gov/docs/ML0719/ML071970394.pdf Accessed August 10, 2018.

Duffin, C. J., and Davidson, J. P., 2011: "Geology and the Dark Side." *Proceedings of the Geologists' Association.* **122**: 7–15.

Dulias, R., and Pelka-Gosciniak, J., 2000: "Aeolian Processes in Different Landscape Zones" (Procesy eoliczne w róznych strefach krajobrazowych). *Dissertations of Faculty of Earth Sciences, University of Silesia.* **5**: 1–196.

Dury, G. H., 1965: *Theoretical Implications of Underfit Streams.* US Geological Survey Professional Paper 452-C. 56 pp.

Dylik, J., 1952: "The Concept of the Periglacial Cycle in Middle Poland." *Bulletin de la société des sciences et des lettres de Lódz.* **3**, 5: 1–29.

Dylik, J., 1964: "L'étude de la Dynamique d'Évolution des Dépressions Fermées à Józéfow aux Environs de Lódz" (The study of the evolutionary dynamics of the closed depressions at Józéfow, near Lódz). *Revue de Géomorphologie Dynamique.* **15**, 10-11-12: Unpaginated.

Dzieduszyńska, D. A., Petera-Zganiacz, J., Murton, J., and Barlow, J., 2023: *Dry Valleys and Dells, Reference Module in Earth Systems and Environmental Sciences.* Elsevier. pp. 1–31.

Eck, P., 1990: *The American Cranberry.* New Brunswick: Rutgers University Press. 420 pp.

Eckhardt, G. M., 1975: *The History of Folsom, New Jersey: 1845–1976.* Bicentennial Edition. Egg Harbor City, NJ: Laureate Press. 121 pp.

Eekhout, J. P. C., Hoitink, A. J. F., and Makaske, B., 2013: "Historical Analysis Indicates Seepage Control on Initiation of Meandering." *Earth Surface Processes and Landforms.* **38**, 8: 888–897.

Ehrenfeld, J. G., 1986: "Wetlands of New Jersey Pine Barrens: the Role of Species Composition in Community Function." *American Midland Naturalist.* **115**, 2: 301–313.

Elmer, L. Q. C., 1869: *History of the Early Settlement and Progress of Cumberland County, New Jersey, and of the Currency of this and Adjoining Colonies.* Bridgeton, NJ: George F.

Nixon. 142 pp.

Engle, C. C., Lee, L. L., and Miller, H. A., 1921: *Soil Survey of the Millville Area, New Jersey*. Geologic Series Bulletin 22, United States Department of Agriculture, Washington, DC. 46 pp.

Ensom, T., Makarieva, O., Morse, P., Kane, D., Aleekseev, V., and Marsh, P., 2020: "The Distribution and Dynamics of Aufeis in Permafrost Regions." *Permafrost and Periglacial Processes*. **31**: 383–395.

Epstein C. M. 1985. "Discovery of the Aquifers of the New Jersey Coastal Plain in the Nineteenth Century." In Epstein C. M., Talkington R. W. (eds.). 1985. *Geological Investigations of the Coastal Plain of Southern New Jersey, part 2: A. Hydrology and the Coastal Plain, B. Paleontologic Investigations*. 2nd Annual Meeting of the Geological Association of New Jersey, Geology Program, Pomona, NJ: Stockton State College. pp. 1–21.

Epstein, C. M., 1997: "A Field Based Hydrologic Classification for Smaller Wetlands." *Wetland Journal*. **9**, 3: 8–11.

Epstein, C. M., 2003: "Is the Pine Barrens Water Table Declining? What Does the Record Show?" In Hozik, M. J., Mihalasky, M. J. (eds.), *Field Guide and Proceedings, 20th Annual Meeting of the Geological Association of New Jersey, October 10–11, 2003*. Trenton, NJ: Geological Association of New Jersey. pp. 79–91.

Evans, L., 1749: *A Map of Pensilvania, New-Jersey, New-York, and the three Delaware counties*. 1 sheet.

Evans, T., 2006: *Moving Out: New Jersey's Population Growth and Migration Patterns*. Trenton, NJ: New Jersey Future. 36 pp.

Everts, and Stewart, 1876: *Combination Atlas Map of Salem and Gloucester Counties, New Jersey*. Philadelphia, PA: Gloucester County Historical Society 123 pp. (Reprinted 2001.)

Everett, T. H., 1981–1982: *The New York Botanical Garden Illustrated Encyclopedia of Horticulture*. New York: Garland Publishing. 3596 pp.

Farr, W. R., 2002: *Waterways of Camden County: A Historical Gazetteer*. Camden, NJ: Camden County Historical Society. 206 pp.

Farr, W. R., 2022. *Watermills of Camden County*. Chapter B, West Jersey History Project, http://www.westjerseyhistory.org/books/farrwatermills/B.shtml Accessed July 24, 2022.

Farr, W. R., 2018. *Watermills of Camden County*. Chapter I, West Jersey History Project, http://www.westjerseyhistory.org/books/farrwatermills/I.shtml Accessed August 10, 2018.

Farrell, S. C., Gagnon, K., Malinousky, T., Columbo, R., Mujica, K., Mitrocsak, J., Cozzi, A., Van Woudenberg, E., and Weisbecker, T., 1985: "Pleistocene? Braided Stream Deposits in the Atsion Quadrangle Area, Northwestern Atlantic County, New Jersey." In Talkington, R. W. (ed.), *Geological Investigations of the Coastal Plain of Southern New Jersey, Part 1: Field Guide. 2nd Annual Meeting of the Geological Association of New Jersey*, Geology Program, Pomona, NJ, Stockton State College. Trenton, NJ: Geological Association of New Jersey. pp. A-1 to A-12.

Faure, H., Walter, R. C., and Grant, D. R., 2002: "The Coastal Oasis: Ice Age Springs on Emerged Continental Shelves." *Global and Planetary Change*. **33**: 47–56.

Fazey, J., Evely, A. C., Reed, M. S., and Stringer, L. C., 2013: "Knowledge Exchange: a Review and Research Agenda for Environmental Management." *Environmental Conservation*. **40**, 1: 19–36.

Fenstermacher, D. E., Rabenhorst, M. C.,

Lang, M. W., McCarty, G. W., and Needelman, B. A., 2014: "Distribution, Morphometry, and Land Use of Delmarva Bays." *Wetlands*. **34**, 6: 1219–1228.

Fenstermacher, D., Rabenhorst, M., Needleman, B. A., Lang, M. W., and McCarty, G. 2015: "Assessing Wetland Morphometrics and Ecosystem Functions in Agricultural Landscapes of the Atlantic Coastal Plain Using Fine Scale Topographic Information." USDA NRCS Technical Notes. https://www.nrcs.usda.gov/Internet/FSE_DOCUMENTS/nrcseprd394811.pdf Accessed August 10, 2018.

Fernow, B., 1877: *Documents Related to the History of the Dutch and Swedish Settlements on the Delaware River, Translated and Compiled from Original Manuscripts in the Office of the Secretary of State, at Albany, and in the Royal Archives, at Stockholm, Volume XII*. Albany, NY: Argus. 669 pp.

FitzPatrick, E. A., 1956: "An Indurated Soil Horizon Formed by Permafrost." *Journal of Soil Science*. **7**, 2: 248–254.

Flemal, R. C., 1972: "Ice Injection Origin of the DeKalb Mounds, North-Central Illinois, U.S.A." In Proceedings, 24th International Geological Congress, Montreal, Section 12. pp. 130–135.

Florea, N., 1970: "Campia cu Crovuri, un Stadiu de Evolutie al Campiilor Loessice." *Studii tehnice si economice*. Seria C, Pedologie, Republica Socialista România, Comitetul Geologic, Institutul Geologic. **16**: 339–353.

Folger, H. P., 1989: "Steelmantown." In Members of the Historical Preservation Society of Upper Township. *A History of Upper Township and Its Villages*. Marmora, NJ: Historical Preservation Society of Upper Township. p. 71.

Forman, R. T. T. (ed.), 1979a: *Pine Barrens: Ecosystem and Landscape*. New York: Academic Press. 601 pp.

Forman, R. T. T., 1979b: "The Pine Barrens of New Jersey: an Ecological Mosaic." In Forman, R. T. T. (ed.), *Pine Barrens: Ecosystem and Landscape*. New York: Academic Press. pp. 569–585.

Fortier, D., Allard, M., and Shur, Y., 2007: "Observation of Rapid Drainage System Development by Thermal Erosion of Ice Wedges on Bylot Island, Canadian Arctic Archipelago." *Permafrost and Periglacial Processes*. **18**: 229–243.

Fothergill, J., 1754: *An Account of the Life and Travels, in the Work of the Ministry, of John Fothergill*. Philadelphia, PA: James Chattin. 280 pp.

Fowler, M., and Herbert, W. A., 1976: *Paper Town of the Pine Barrens: Harrisville, New Jersey*. Eatontown, NJ: Environmental Education Publishing Service. 52 pp.

Fowler, T. M., and Lummis, 1880: *Map of Merton Tract in Hamilton Township—Atlantic Co. N.J. Scale 20 ch's to inch, 18,062 acres*.

French, H. M., 1972: "Asymmetrical Slope Development in the Chiltern Hills." *Biuletyn Peryglacjalny*. **21**: 51–73.

French, H. M., 1996: *The Periglacial Environment*. 2nd edition. Harlow, Essex: Addison Wesley Longman. 341 pp.

French, H. M., 2000: "Does Lozinski's Periglacial Realm Exist Today? A Discussion Relevant to Modern Usage of the Term 'periglacial.'" *Permafrost and Periglacial Processes*. **11**: 35–42.

French, H. M., 2002: "Thaw vs. Melt: an editorial." *Frozen Ground*. **26**: 6.

French, H. M., 2003: The Development of Periglacial Geomorphology: 1– up to 1965." *Permafrost and Periglacial Processes*. **14**: 29–60.

French, H. M., 2007: *The Periglacial Environment* (3rd edition). Chichester, UK: John Wiley & Sons. 458 pp.

French, H. M., 2018: *The Periglacial Environment* (4th edition). Hoboken, NJ: John Wiley & Sons. 544 pp.

French, H. M., and Demitroff, M., 2001: "Cold-Climate Origin of the Enclosed Depressions and Wetlands ('Spungs') of the Pine Barrens, Southern New Jersey, USA." *Permafrost and Periglacial Processes.* **12**: 337–350.

French, H. M., and Demitroff, M., 2003: "Late Pleistocene Periglacial Phenomena in the Pine Barrens of Southern New Jersey." GANJ Field Excursion Guide, October 11, 2003. In Hozik, M. J., Mihalasky, M. J. (eds.), *Field Guide and Proceedings, 20th Annual Meeting of the Geological Association of New Jersey, October 10–11, 2003.* Trenton, NJ: Geological Association of New Jersey. pp. 117–142.

French, H. M., and Demitroff, M., 2012: "Late-Pleistocene Paleohydrology, Eolian Activity and Frozen Ground, New Jersey Pine Barrens, Eastern USA." *Netherlands Journal of Geosciences.* **91**, 1/2: 25–35

French, H. M., Demitroff, M., and Forman, S. L., 2003: "Evidence for Late-Pleistocene Permafrost in the New Jersey Pine Barrens (Latitude 39°N), Eastern USA." *Permafrost and Periglacial Processes.* **14**: 259–274.

French, H. M., Demitroff, M., and Forman, S. L., 2005: "Evidence for Late-Pleistocene Thermokarst in the New Jersey Pine Barrens (Latitude 39°N), Eastern USA." *Permafrost and Periglacial Processes.* **16**: 173–186.

French, H. M., Demitroff, M., Forman, S. L., and Newell, W. L., 2007: "A Chronology of Late-Pleistocene Permafrost Events in Southern New Jersey, Eastern USA." *Permafrost and Periglacial Processes.* **18**: 49–59.

French, H. M., Demitroff, M., and Newell, W. L., 2009a: "Past Permafrost on the Mid-Atlantic Coastal Plain, Eastern United States. *Permafrost and Periglacial Processes.* **20**, 3: 285–294.

French, H. M., Demitroff, M., Streletskiy, D., Forman, S. L., Gozdzik, J., Konishchev, V. N., Rogov, V. V., and Lebedeva-Verba, M. P., 2009b: "Evidence for Late-Pleistocene Permafrost in the Pine Barrens, Southern New Jersey." *Earth's Cryosphere.* **2009**, 3: 17–28 (in Russian).

French, H. M., and Millar, S. W. S., 2014: "Permafrost at the Time of the Last Glacial Maximum (LGM) in North America." *Boreas.* **43**, 3: 667–677.

French, H. M., and Thorn, C. E., 2006: "The Changing Nature of Periglacial Geomorphology." *Géomorphologie: Relief, Processus, Environment.* **3**: 165–174.

Frenzel, B. (ed.), 1995: "European River Activity and Climatic Change during the Lateglacial and Early Holocene." Special Issue: ESF Project "European Paleoclimate and Man" 9. *Paläoklimaforschung/Paleoclimate Research.* **14**: 1–226.

Fritz, J. M., 2012: "The Carmans of Carmantown, Pleasantville and Atlantic City." *Yearbook with Historical and Genealogical Journal*, The Atlantic County Historical Society. **17**, 1: 27–38.

Frodeman, R., 1995: "Geological Reasoning: Geology as an Interpretive and Historical Science." *Geological Society of America Bulletin.* **107**, 8: 960–968.

Froehlich, W, and Slupik, J., 1982: "River Icings and Fluvial Activity in Extreme Continental Climate: Khangai Mountains, Mongolia." In French, H. M. (ed.), *The Roger J. E. Brown Memorial Volume. Proceedings of the Fourth Canadian Permafrost Conference, Calgary, Alberta, March 2–6, 1981.* Ottawa: National Research Council of Canada, pp. 203–211.

Fry, M., 1916: "Laying out Landis Avenue." *Vineland Historical Magazine.* **1**: 2.

Fry, M., 1918: "A Tar Kiln." *Vineland*

Historical Magazine. **3**: 13.

Fryberger, S. G., Schenk, C. J., and Krystinik, L. F., 1988: "Stokes Surfaces and the Effects of Near-Surface Groundwater-Table on Aeolian Deposition." *Sedimentology.* **35**: 21–41.

Fuller, S., 2005: *Kuhn vs. Popper: The Struggle for the Soul of Science.* New York: Columbia University Press. 160 pp.

Gao, C., 2014: "Relict Thermal-Contraction-Crack Polygons and Past Permafrost South of the Late Wisconsinan Glacial Limit in the Mid-Atlantic Coastal Plain, USA." Short communication. *Permafrost and Periglacial Processes.* **25**: 144–149.

Gębica, P., Michczyńska, D. J., and Starkel, L., 2015: "Fluvial History of the Sub-Carpathian Basins (Poland) during the Last Cold Stage (60–8 cal ka BP)." *Quaternary International.* 388: 119–141.

Geubtner, E., 2008. *Buena Vista Township—Railroad Revitalization Project.* Memo letter from Ed Geubtner to John Lamey, Mullin & Lonergan Associates, Camp Hill, PA. May 20, 2008.

Gherghina, A., Grecu, F., and Molin, P., 2008: "Morphometrical Analysis of Microdepressions in the Central Baragan Plain (Romania)." *Revista de Geomorfologie.* **10**: 31–38.

Gibson, D. J., Zampella, R. A., and Windisch, A. G., 1999: "New Jersey Pine Plains: the 'true barrens' of the New Jersey Pine Barrens." In Anderson, R. C., Fralish, J. S., and Baskin, J. M. (eds.), *Savannas, Barrens, and Rock Outcrop Plant Communities of North America.* Cambridge: Cambridge University Press. pp. 52–66.

Gifford, J., 1896: "Report on Forest Fires for Season of 1895." In Geological Survey of New Jersey. *Annual Report of the State Geologist for 1895.* Trenton, NJ: John L. Murphy. pp. 157–175.

Gifford, J., 1900: "The Forestal Conditions and Silvicultural Prospects of the Coastal Plain of New Jersey." In Geological Survey of New Jersey. *Annual Report of the State Geologist for the Year 1889: Report on Forests.* Trenton, NJ: MacCrellish & Quigley. pp. 233–319.

Gignac, L. D., Halsey, L. A., and Vitt, D. H., 2000: "A Bioclimatic Model for the Distribution of *Sphagnum*-Dominated Peatlands in North America under Present Climatic Conditions." *Journal of Biogeography.* **27**: 1139–1151.

Gillijins, K., Poesen, J., and Deckers, J., 2005: "On the Characteristics and Origin of Closed Depressions in Loess-Derived Soils in Europe—a Case Study from Central Belgium." *Catena.* **60**: 43–58.

Gladden, D., 2011: "Staging the Trauma of the Bog in Marina Carr's By the Bog of Cats..." *Irish Studies Review.* **19**, 4: 387–400.

Gloucester County Road Return, c. 1790s: Book A, page 175.

Godin, E., Fortier, D., and Coulombe, S., 2014: "Effects of Thermo-Erosion Gullying on Hydrologic Flow Networks, Discharge and Soil Loss." *Environmental Research Letters.* **9**, 10: 1–10.

Goertzel, T., and Leonardis, J., 2006: "Public Opinion on Sprawl and Smart Growth in Southern New Jersey." Sociology Department, Rutgers University, Camden, NJ. http://crab.rutgers.edu/~goertzel/sjgrowth.htm. Accessed August 10, 2018.

Gonera, P., 1986: "Zmiany Geometrii Koryt Meandrowych Warty na tle Wahan Klimatycznych w Póznym Vistulianie i Holocenie" (Changes in geometry of the Warta meandering channels against climatic fluctuations during the Late Vistulian and Holocene). *Wydawnictwo Naukowe Uniwersytetu im Adama Mickiewicza w*

Poznaniu, Seria Geografia. **33**: 1–84.

Good, R. E., Good, N. F., and Andresen, J. W., 1979: "The Pine Barren Plains." In Forman, R. T. T. (ed.), *Pine Barrens: Ecosystem and Landscape.* New York: Academic Press. pp. 283–295.

Gordon, T., 1828: *A Map of the State of New Jersey with Part of the Adjoining States Compiled under Patronage of the Legislature of Said State.* Philadelphia, PA: Thomas Gordon. 1 sheet.

Gordon, T., 1850: *A Map of New Jersey with Part of the Adjoining States.* H.S. Tanner, Philadelphia, Pennsylvania, 1 sheet. (Revised, corrected and improved by Robert T. Horner.)

Gordon, T., 2015: "Herbert Payne: Last of the Old-Time Charcoal Makers and His Coaling Process in the Pine Barrens of New Jersey." Galloway, NJ: South Jersey Culture & History Center. 12 pp.

Gordon, T., and Demitroff, M., 2009: "13 August: Spungs, Cripples, Blue Holes, and Savannahs (Savannas) in the Pine Barrens of Atlantic and Cumberland Counties, NJ." *Bartonia: Journal of the Philadelphia Botanical Club.* **64**: 59–62.

Gould, S. J., 1986: "Evolution and the Triumph of Homology, or Why History Matters." *American Scientist.* **74**: 60–69.

Grace's Guide, 2018: "Spong and Co. Grace's Guide to British Industrial History." https://www.gracesguide.co.uk/Spong_and_Co Accessed February 14, 2021.

Graham, T. B., 2016: "Climate Change and Ephemeral Pool Ecosystems: Potholes and Vernal Pools as Potential Indicator Systems." Impacts of Climate Change on Land and Ecosystems, US Geological Survey, http://geochange.er.usgs.gov/sw/impacts/biology/vernal/ Accessed August 11, 2018.

Gray, M., 2013: *Geodiversity: Valuing and Conserving Abiotic Nature* (2nd edition). Chichester, UK: Wiley-Blackwell. 512 pp.

Greeley, R., and Iversen, J. D., 1985: *Wind as a Geological Process on Earth, Mars, Venus and Titan.* Cambridge Planetary Sciences Series 4. Cambridge University Press: Cambridge. 333 pp.

Green, C. F., ca. 1920s?: *A Place of Olden Days, Nescochague, Sweetwater, Pleasant Mills: a Historical Sketch.* Hammonton: Hammonton Printing. 32 pp.

Green, S. A., 1879: *An Account of the Early Land-Grants of Groton, Massachusetts.* Cambridge, MA: University Press, John Wilson and Son. 56 pp.

Green, S. A., 1894: *An Historical Sketch of the Town of Groton.* (Reprinted from "The History of Middlesex County, Massachusetts." Groton, MA: s.n. 263 pp.

Green, S. A., 1912: *The Natural History and the Topography of Groton, Massachusetts, Together with Other Matter Relating to the History of the Town.* Massachusetts. Cambridge, MA: University Press, John Wilson and Son. 207 pp.

Grimwood, B., Reddy, M., and Goodwin, G., 2004: "Particular notes." *Common Ground.* **3**: unpaginated, 4 pp.

Groot, J. J., Benson, R. N., and Wehmiller, J. F., 1995: "Palynological, Foraminiferal and Aminostratigraphic Studies of Quaternary Sediments from the U.S. Middle Atlantic Upper Continental Shelf and Coastal Plain." *Quaternary Science Review.* **14**: 17–49.

Grootjans, A. P., Bakker, J. P., Jansen, A. J. M., and Kemmers, R. H., 2002: "Restoration of Brook Valley Meadows in the Netherlands." *Hydrobiologia.* **478**: 149–170.

Gross, E., Merritts, D., Walter, R., Demitroff, M., and Huot, S., 2017: "Late Pleistocene Paleoclimate Inferred from Sediment Infill in a Thermal Contraction-Wedge in Shale Bedrock near Carlisle, Pennsylvania." 6th

Annual Amtrak Club Meeting Penn State University, Abstracts Volume: 15.

Grumet, R. S., 2014: *Beyond Manhattan: Gazetteer of Delaware Indian History Reflected in Modern-Day Place Names.* New York State Museum Record 5. Albany, NY: The New York State Museum New York. 242 pp. + addendum.

Grundmann, R., 2017: "The Problem of Expertise in Knowledge Societies." *Minerva, A Review of Science, Learning and Policy.* **55**, 1: 25–48.

Gubin, S. V., and Lupachev, A. V., 2008: "Soil Formation and the Underlying Permafrost." *Eurasian Soil Science.* **41**, 6: 574–585.

Gustavson, T. C., Holliday, V. T., and Hovorka, S. D., 1995: "Origin and Development of Playa Basins, Sources of Recharge to the Ogallala Aquifer, Southern High Plains, Texas and New Mexico." Report of Investigation No.229, Bureau of Economic Geology, The University of Texas at Austin. 44 pp.

Haas, P. M., 2004: "When Does Power Listen to Truth? A Constructivist Approach to the Policy Process." *Journal of European Public Policy.* **11**, 4: 569–592.

Hack, J. T., 1953: "Geologic Evidence of Late Pleistocene Climates." In Shapely, H. (ed.), *Climate Change: Evidence, Causes, and Effects.* Cambridge: Harvard. pp. 179–188.

Hackney, J., 1795: Survey for William Hollingshead, 19,623 acres of land situate part in the Township of Great Egg Harbour in the County of Gloucester and the remaining part in the Township of Prince Morris River in the County of Cumberland. Certified by John Clement in April 17th, 1832. Vineland Historical Society. 5 sheets.

Haffner, J., and Mitchell, K., 2019: "The Palisades in Peril: Meet the Forgotten Women Who Fought to Save the Great Cliffs of the Hudson River." Behind the Scenes, New York Historical Society Museum & Library. https://behindthescenes.nyhistory.org/palisades-peril-forgotten-women-saved/ Accessed February 06, 2021.

Hajna, L. R., 2000a: "Ponds in Pinelands Drying Up." *Courier Post: South Jersey's Newspaper.* September 05: 1A, 8A.

Hajna, L. R., 2000b: "Pinelands Shrinking Ponds Have State Officials Puzzled." *Courier Post: South Jersey's Newspaper.* September 15: 1B, 4B.

Hajna, L. R., 2002: "Drought/Thousands of Pools Have Dried Up across the State." *Courier Post: South Jersey's Newspaper.* March 20: 1A, 6A.

Hajna, L. R., 2004: "Are the Pinelands Drying Up?" *Courier Post: South Jersey's Newspaper.* March 28: 10A.

Haldorsen, S., and Treidel, H., 2012: "Palaeogroundwater Dynamics and Their Importance for Past Human Settlements and Today's Water Management." *Quaternary International.* **257**: 1–3.

Hall, C. J., 1942: "Charles Dexter McFarlin of South Carver, Massachusetts Carried Cranberry Cultivation to Oregon." *Cranberries: The National Cranberry Magazine.* **8**, 2, June: 6.

Hall, J. H. (chief ed.), 2012: *Dictionary of American Regional English, Volume V: Sl–Z.* Cambridge, MA: Belknap Press of Harvard University Press. 1296 pp.

Hall, K., 1997: "Zoological Erosion in Permafrost Environments: a Possible Origin of Dells?" *Polar Geography.* **21**, 1: 1–9.

Hambrey, M. J., and Fitzsimons, S. J., 2010: "Development of Sediment-Landform Associations at Cold Glacier Margins, Dry Valleys, Antarctica." *Sedimentology.* **57**: 857–882.

Hamelin, Louis-Edmond, and Cook, F. A., 1967: *Illustrated Glossary of Periglacial Phenomena.* Québec: Les Presses de l'Université Laval. 237 pp.

Hamilton, P. B., Siver, P. A., and Reimer, C. W., 2004: *Neidium rudimentarum* (Bacillariophyceae), a rare species with novel rudimentary canals and areolar openings from the temperate region of North America. *Rhodora*. **106**: 33–42.

Hampton, G., 1917: The "Old Cohansey Road." Letter from Hon. John A. Ackley to Ex-Mayor George Hampton, Bridgeton, N.J., April 3, 1917. *The Vineland Historical Magazine*. **2**: 49–50.

Harman, J. R., and Arbogast, A. F., 2004: "Environmental Ethics and Coastal Dunes in Western Lower Michigan: Developing a Rationale for Ecosystem Preservation." *Annals of the Association of American Geographers*. **94**, 1: 23–36.

Harmar, J., Armstrong, E., and Denny, E. (eds.), 1860: *The Record of the Court at Upland, in Pennsylvania. 1676-1681. And a Military Journal Kept by Major E. Denny, 1781-1795*. Philadelphia, PA: J. B. Lippincott. 498 pp.

Harper, D., 1950: unpublished Senior thesis, Princeton University, Princeton, NJ.

Harper, D., 2004: "Richland Man's Study is Dry Work." *The Press of Atlantic City*. June 21: C1, C4.

Harper, R. M., 1914: "The Coniferous Forests of Eastern North America." *Popular Science Monthly*. **85**: 338–361.

Harris, S., 2004: "Possible Multiple Origins of the Younger Dryas 'Viviers' and Other Permafrost-Related Features on the Hautes Fagnes in Belgium." *Geophysical Research Abstracts*. **6**: 01253, 2004. 2 pp.

Harshberger, J. W., 1916: *The Vegetation of the New Jersey Pine Barrens: An Ecological Investigation*. Philadelphia, PA: Christopher Sower. 329 pp.

Hart, J. F., 1964: "Land and Life-Selected Writings of Carl Sauer" (book review). *Annals of the Association of American Geographers*. **54**, 4: 612–614.

Hartman, C. S., 1950: Map compiled by Chas. S. Hartman of Port Elizabeth, NJ. Byerley & Dorchester surveys recorded Burlington, NJ. Lib. M, Folio 15. Survey made in Apr., May & June 1691. 1 sheet.

Hartman, C. S., 1978: Eight maps made by Charles S. Hartman, 207 W. Main Street, Millville, NJ, during period 1920 and 1978 and were compiled from searches made at Courthouses Bridgeton, Salem, Cape May, Mays Landing, Woodbury, Gloucester and from ancient survey maps made by old surveyors. 20 chains = 1 inch. Single sheet.

Hartzog (Bierbrauer), S., 1982: "Palynology and Late Pleistocene-Holocene Environment on the New Jersey Coastal Plain." In Sinton, J. W. (ed.), *History, Culture and Archeology of the New Jersey Pine Barrens*. Center for Environmental Research, Pomona, NJ: Stockton State College. pp. 6–14.

Harvey, D., 1969: *Explanation in Geography*. New York: St. Martin's Press. 521 pp.

Hayfield, C., and Brough, M., 1986: "Dewponds and Pondmakers of the Yorkshire Wolds." *Folk Life*, Journal of Ethnological Studies. **25**, 1: 74–91.

Haynes, C. V. Jr., and Agogino, G. A., 1966: "Prehistoric Springs and Geochronology of the Clovis Site, New Mexico." *American Antiquity*. **31**, 6: 812–821.

Head, M. J., 2019: "Formal Subdivision of the Quaternary System/Period: Present Status and Future Directions." *Quaternary International*. **500**: 32–51.

Herschy, R. W., 2012: "Dew Ponds." In Bengtsson, L., Herschy, R. W., Fairbridge, R. W. (eds.), *Encyclopedia of Lakes and Reservoirs*. Dordrecht, DE: Springer. pp. 217–219.

Hess, T., 1910: *Jonathan and Hannah Steelman: A Family Tree or Record of the Decedents of their Four Children.*

Millville, NJ: self-published. 106 pp.

Heston, A. M., 1924: *South Jersey: A History, 1664–1924.* New York: Lewis Historical Publishing. 4 volumes.

Hider, J., 1745: Survey and map of 100 acres at a place known as The Ponds, West Jersey Loose Records 1745, 40648.

Hilton, A., 1999: "In Search of a Legend . . . Tales of Local Haunts Revived in Williamstown." *Plain Dealer*, July 15: Unpaginated.

Holland, S., 1776: "The Provinces of New York and New Jersey; With Part of Pensilvania, and the Governments of Trois Riviers, and Montral, Drawn by Capt. Holland." *In* Jefferys, T., *The American Atlas: A Geographical Description of the Whole Continent of America.* London: Sayer R., and Bennett, J., Plate no. 17 of 29.

Hollinshead, T., 1779: A survey of 85.5 acres at a place called Eryens Barrons, Gloucester Township, Gloucester County, NJ. West Jersey Loose Records, 1779, 46023.

Hopkins, D. M, Karlstrom, T. N. V., Black, R. F., Williams, J. R., Péwé, T. L., Fernold, A. T., and Muller, E. H., 1955: "Permafrost and Groundwater in Alaska." US Geological Survey Professional Paper 264-F. pp. 113–146.

Hopkins, G. M., 1873: *Combined Atlas of the State of New Jersey and the City of Newark.* Newark, NJ: G. M. Hopkins & Co. 120 pp.

Hopping, W. H., 1886: *Reminiscences of Atlantic County: Mays Landing.* Hammonton, NJ: Mirror Job and Book Printing. 39 pp.

Hubbard, A. J., and Hubbard, G., 1907: *Neolithic Dew-Ponds and Cattle-Ways.* 2nd edition. London: Longmans, Green. 116 pp.

Hufford, M., 1986: *One Space, Many Places: Folklife and Land Use in New Jersey's Pinelands National Reserve.* Washington, DC: American Folklife Center. 144 pp.

Huijzer, A. S., 1993: *Cryogenic Microfabrics and Macrostructures: Interrelations, Processes, and Paleoenvironmental Significance.* Ph.D. dissertation, Amsterdam: Vrije Universiteit, 245 pp.

Huisink, M., 2000: "Changing River Styles in Response to Weichselian Climate Changes in the Vecht Valley, Eastern Netherlands." *Sedimentary Geology.* **133**: 115–134.

Humlum, O., 1979: "Icing Ridges: a Sedimentary Criterion for Recognizing Former Occurrence of Icings." *Bulletin of the Geological Society of Denmark.* **28**: 11–16.

Hummel, J., 2015: Unpublished George Agnew Chamberlain short story "'Damiana' under wraps." NJ.com, Cumberland County, February 09, 2015, https://www.nj.com/cumberland/index.ssf/2015/02/george_agnew_chamberlain_short_story_under_wraps.html Accessed August 09, 2018.

Hutchings, M. C., 1907a: "Historical Sketches of Buena," by the editor, Chapter IV. *Valley Ventura*, October 5.

Hutchings, M. C., 1907b: "Historical Sketches of Buena," by the editor, Chapter V. *Valley Ventura*, October 19.

Hutchinson, J. N., 1980: "Possible Late Quaternary Pingo Remnants in Central London." *Nature.* **284**: 253–255.

I., 1832: "The Variety of Game in New Jersey." *In* Doughty, J., and Doughty, T., *The Cabinet of Natural History and American Rural Sports*, Vol. 2, January 4, pp. 15–18.

Ingrouille, M., 1995: "The Managed Landscape: Fields, Pastures, Woods and Gardens." *In* Ingrouille, M. *Historical Ecology of the British Flora.* Dordrecht: Springer Netherlands, pp. 205–311.

Ivester, A. H., and Leigh, D. S., 2003: "Riverine Dunes on the Coastal Plain of Georgia, USA." *Geomorphology*. **51**: 289–311.

Ivester, A. H., Leigh, D. S., and Godfrey-Smith, D. I ., 2001: "Chronology of Inland Eolian Dunes on the Coastal Plain of Georgia, USA." *Quaternary Research*. **55**: 293–302.

Jahn, A., 1956: Wyżyna Lubelska. "Rzeźba i Czwartorzęd" (Geomorphology and Quaternary history of Lublin Plateau). *Polska Akademia Nauk Instytut Geografii, Prace Geograficzne*. 7. Warszawa, POL: Panstwowe Wydawnictwo Naukowe. 453 pp.

Jahn, A., 1975: *Problems of the Periglacial Zone (Zagadnienia strefy peryglacjalnej)*. Translated from Polish. Warszawa, POL: Polish Scientific Publishers. 223 pp.

Jahn, R., 1980: *Down Barnegat Bay: A Nor'easter Midnight Reader*. Mantoloking, NJ: Beachcomber Press. 207 pp.

Jambert, C., Delmas, R., Serca, D., Thouron, L., Labroue, L., and Delprat, L., 1997: "N_2O and CH_4 Emissions from Fertilized Agricultural Soils in Southwest France." *Nutrient Cycling in Agroecosystems*. **48**: 105–114.

Johnson, D., 1942: *The Origin of the Carolina Bays*. New York: Columbia University Press. 341 pp

Johnson, R. F., 2001: *Weymouth, New Jersey: A History of the Furnace, Forge, and Paper Mills*. Kearney, NE: Morris Publishing. 137 pp.

Johnson, W., 1908: *Folk Memory or the Continuity of British Archaeology*. Oxford, UK: Clarendon Press. 416 pp.

Johnsson, G., 1961: "Periglaciala Dalar i Sydligaste Sverige" (Periglacial valleys in southernmost Sweden). *Geologiska Föreningens Förhandlingar*. **83**, 2: 162–185.

Jones, D., 1949: Comments and/or questions raised by the conference. In Assembled notes from the "Friends" Pensauken field trip, May. pp. 4–5. Notes on file at the NJ Geological Survey, Trenton, NJ.

Jones, M. D., 2013: "What Do We Mean by Wet? Geoarchaeology and the Reconstruction of Water Availability." *Quaternary International*. **308–309**: 76–79.

Jungerius, P. D., 1984: "A Simulation Model of Blowout Development." *Earth Surface Processes and Landforms*. **9**: 509–512.

Jungerius, P. D., and van der Meulen, F., 1989: "The Development of Dune Blowouts, as Measured with Erosion Pins and Sequential Air Photos." *Catena*. **16**: 369–376.

Juelg, G. R., 2003: *New Jersey Pinelands Threatened and Endangered Species*. 2nd ed. Southampton, NJ Pinelands Preservation Alliance. 88 pp.

Juelg, G. R., 2014: "Remarkable Ponds of the Pinelands." *Inside the Pinelands* **21**, 4, June/July: 6.

Kaczorek, D., Sommer, M., Andruschkewitsch, I., Oktaba, L., Czerwinski, Z., and Stahr, K., 2004: "A Comparative Micromorphological and Chemical Study of 'raseneisenstein' (bog iron ore) and 'ortstein.'" *Geoderma*. **121**: 83–94.

Kaczorowski, R. T., 1977: *The Carolina Bays: A Comparison with Modern Oriented Lakes*. Technical Report No. 13-CRD, Coastal Research Division, University of South Carolina. 124 pp.

Kalm, P., 1770: *Peter Kalm's Travels in North America—The America of 1750* (2 volumes), revised and adapted from the original Swedish and edited by Adolph B. Benson). New York: Dover Publications. 797 pp. (Reprinted 1964).

Kaskey, J., 2000: "As Pinelands Ponds Dry, History Goes with Them." *The Press of Atlantic City*. **241**, August 28, MM. Pleasantville, NJ: South Jersey Publishing Co. pp. A1, A7.

Kasse, C., 1998: "Depositional Model for Cold-Climate Tundra Rivers." In

Benito, G., Baker, V. R., and Gregory, K. J. (eds.), *Palaeohydrology and Environmental Change*. Chichester, UK: John Wiley & Sons. pp. 83–97.

Kasse, C., Bohncke, S. J. P., Vandenberghe, J., and Gabris, G., 2010: "Fluvial Style Changes During the Last Glacial-Interglacial Transition in the Middle Tisza Valley (Hungary)." *Proceedings of the Geologists' Association*. **121**: 180–194.

Kasse, C., Vandenberghe, J., van Huissteden, J., Bohncke, S. J. P., and Bos, J. A. A., 2003: "Sensitivity of Weichselian Fluvial Systems to Climate Change (Nochten mine, eastern Germany)." *Quaternary Science Reviews*. **22**: 2141–2156

Kates, R. W., 2011: "What Kind of a Science is Sustainability Science?" *Proceedings of the National Academy of Science*. **108**, 49: 19449–19450.

Kavetski, D., and Clark, M. P., 2010: "Ancient Numerical Daemons of Conceptual Hydrological Modeling: 2. Impact of Time Stepping Schemes on Model Analysis." *Water Resources Research*. **46**, W10511.

Kecskes, R., 2014: "Water is Not Unlimited in the Pines." *Inside the Pinelands* **21,** 4, June/July: 3.

Kelly, J. F., 2018: "Status Surveys for the Endangered Plant Species *Narthecium americanum* (Bog Asphodel, Liliaceae) in the New Jersey Pine Barrens, Using Remote Sensing and Geographic Information Systems." *Journal of the Torrey Botanical Society*. **145**: 69–81.

Kenney, L. P., and Burne, M. R., 2001: *A Field Guide to the Animals of Vernal Pools*. Westborough, MA: Massachusetts Division of Fisheries and Wildlife. 78 pp.

Kerr, W. C., 1881: "On the Action of Frost in the Arrangement of Superficial Earthy Material." *American Journal of Science*. 3rd Series. **21**: 345–359.

Khan, L., 1872: *One day, Levin . . .*
He be Free—William Still and the Underground Railroad. New York: Dutton. 231 pp. (Reprinted 1972).

Kinsey, D. N., 2008: "Has the Mount Laurel Doctrine Delivered on Smart Growth?" *Planning & Environmental Law*. **60**, 6: 3–9.

Kiss, T., Hernesz, P., Sümeghy, B., Györgyövics, K., and Sipos, G., 2015: "The Evolution of the Great Hungarian Plain Fluvial System—Fluvial Processes in a Subsiding Area from the Beginning of the Weichselian." *Quaternary International*. **388**: 142–155.

Klatkowa, H., 1965: "Niecki i Doliny Denudacyjne w Okolicach Lódzi" (Vallons en berceau et vallees seches aux environs de Lódz / Small cradled valleys and dry valleys around Lódz). *Acta Geographica Lodziensia*. **19**: 1–144.

Klempe, H., 2015: "The Hydrogeological and Cultural Background for Two Sacred Springs, Bø, Telemark County, Norway." *Quaternary International*. **368**: 31–42.

Klinger, L. F., 1996: "The Myth of the Classic Hydrosere Model of Bog Succession." *Arctic and Alpine Research*. **28**, 1: 1–9.

Knight, A. T., Cowling, R. M., Rouget, M., Balmford, A., Lombard, A. T., and Campbell, B. M., 2008: "Knowing But Not Doing: Selecting Priority Conservation Areas and the Research-Implementation Gap." *Conservation Biology*. **22**, 3: 610–617.

Konikow, L. F., 2013: *Groundwater depletion in the United States (1900-2008)*. US Geological Survey Scientific Investigations Report 2013-5079, 63 p.

Konishchev, V. N., 1982: "Characteristics of Cryogenic Weathering in the Permafrost Zone of the European USSR." *Arctic and Alpine Research*. **14**, 3: 261–265.

Konishchev, V. N., 1997: "A Cryo-lithogenic Method for Estimating

Palaeotemperature Conditions During Formation of the Ice Complex and Subaerial Periglacial Sediments." *Earth's Cryosphere.* **1**, 2: 23–28.

Konishchev, V. N., Lebedeva-Verba, M. P., Rogov, V. V., and Stalina, E. E., 2005: *Cryogenesis of Modern and Late Pleistocene Deposits of the Altai and the Periglacial Region of Europe* (in Russian). Moscow, RUS: GEOS. 133 pp.

Konishchev, V. N., and Rogov, V. V., 1993: "Investigations of Cryogenic Weathering in Europe and Northern Asia." *Permafrost and Periglacial Processes.* **4**: 49–64.

Konishchev, V. N., and Rogov, V. V., 2017: "Phenomena of the Processes of Cryogenesis in Loess Composition." *Geography, Environment, Sustainability* (GES Journal), **10**, 2: 4–14.

Koster, E. A., 2009: "The 'European Aeolian Sand Belt': Geoconservation of Drift Sand Landscapes." *Geoheritage.* **1**: 93–110.

Kozarski, S., 1975: "Oriented Kettle Hole in Outwash Plains." *Quaestiones Geographicae.* **2**: 99–112.

Kozarski, S. (ed.), 1991: "Late Vistilian (=Weichselian) and Holocene Aeolian Phenomena in Central and Northern Europe." *Zeitschrift für Geomorphologie,* Supplementbände 90: 1–270.

Kozarski, S., 1995: "The Periglacial Impact on the Deglaciated Area of Northern Poland after 20-kyr bp." *Biuletyn Peryglacjalny.* **34**: 73–102.

Kuchukov, E. Z., and Malinovsky, D. V., 1983: "Erodibility of Unconsolidated Materials in Permafrost." In *Fourth International Conference on Permafrost, Proceedings, Volume 1.* Washington, DC: National Academy Press. pp. 672–676.

Lacombe, P. J., Carleton, G. B., Pope, D. A., and Rice, D. E., 2009: "Future Water-Supply Scenarios, Cape May County, New Jersey, 2003–2050." US Geological Survey Scientific Investigations Report, 2009-5187, 158 pp.

Ladd, J., 1748: Survey and Map of 250 Acres at a Place Called the Indian Orchard. West Jersey Loose Records, 1748, 40813.

Laidig, K. J., 2012: "Simulating the Effect of Groundwater Withdrawals on Intermittent-Pond Vegetation Communities." *Ecohydrology.* **5**: 841–852.

Laidig, K. J., Zampella, R. A., Brown, A. M., and Procopio, N. A., 2010: "Development of Vegetation Models to Predict the Potential Effect of Groundwater Withdrawals on Forested Wetlands." *Wetlands.* **30**: 489–500.

Laidig, K. J., Zampella, R. A., Bunnell, J. F., Dow, C. L., and Sulikowski, T. M., 2001: *Characteristics of Selected Pine Barrens Treefrog Ponds in the New Jersey Pinelands.* Long-term Environmental-Monitoring Program. Pinelands Commission, New Lisbon, NJ. 43 pp.

Laity, J. E., 1994: "Landforms of Eolian Erosion." In Abrahams, A. D., Parsons, A. J. (eds.), *Geomorphology of Desert Environments.* London: Chapman & Hall. 674 pp.

Landis, C. K., 1916: "Journal of Charles K. Landis." *Vineland Historical Magazine.* **1**, 3: 36–41.

Landscape Design Associates for the Countryside Commission, 1996: *The Landscape of the Sussex Downs: Area of Outstanding Natural Beauty.* Storrington, UK: Sussex Downs Conservation Board. 88 pp.

Lang, A., and Mauz, B., 2006: "Towards Chronologies of Gully Formation: Optical Dating of Gully Fill Sediments from Central Europe." *Quaternary Science Reviews.* **25**: 2666–2675.

Langohr, R., and Sanders, J., 1985: "The

Belgium Loess belt in the Last 20,000 Years: Evolution of Soils and Relief in the Zonien Forest." *In* Boardman, J. (ed.), *Soils and Quaternary Landscape Evolution.* Chichester, UK: John Wiley and Sons. 359–371.

Lantis, J., 1949: My personal, temporary interpretation of the origin of the Pensauken sands and gravels is as follows. *In* Assembled notes from the "Friends" Pensauken field trip, May. pp. 9–10. Notes on file at the NJ Geological Survey, Trenton, NJ.

Lathrop, R. G., Montesano, P., Tesauro, J., and Zarate, B., 2005: "Statewide Mapping and Assessment of Vernal Pools: A New Jersey Case Study. *Journal of Environmental Management.* **76**: 230–238.

Lattanzi, G. D., 2017: "Current Research on Paleoindian Occupations in New Jersey." *Journal of Middle Atlantic Archaeology.* **33**: 49–61.

Leap, B., 1996: "The Legend of the Bottomless Blue Hole." *South Jersey Magazine.* **25**, 4: 45–47 (reprinted from *Camden County Historical Society Newsletter.* Winter, 1982: 3).

Lee, F. B., 1896: "Jerseyisms." *Dialect Notes.* **1**: 327–337 (reprinted in *SoJourn: A Journal Devoted to the History, Culture, and Geography of South Jersey*, **2**, 1: 59–67).

Leigh, D. S., 2006: "Terminal Pleistocene Braided to Meandering Transition in Rivers of the Southeastern USA." *Catena.* **66**: 155–160.

Leigh, D. S., 2008: "Late Quaternary Climates and River Channels of the Atlantic Coastal Plain, Southeastern USA." *Geomorphology.* **101**: 90–108.

Leigh, D. S., Srivastava, P., and Brook, G. A., 2004: "Late Pleistocene Braided Rivers of the Atlantic Coastal Plain, USA." *Quaternary Science Reviews.* **23**: 65–84.

Lemcke, M. D., and Nelson, F. E., 2003: "Periglacial Sediment-Filled Wedges, Northern Delaware, USA." *In* Hozik, M. J., Mihalasky, M. J. (eds.), *Field Guide and Proceedings, 20th Annual Meeting of the Geological Association of New Jersey, October 10–11, 2003.* Trenton, NJ: Geological Society of New Jersey. pp. 93–99.

Lemcke, M. D., and Nelson, F. E., 2004: "Cryogenic Sediment-Filled Wedges, Northern Delaware, USA. *Permafrost and Periglacial Processes.* **15**: 319–326.

Levins, H., 2008: "Finding the World's First Dinosaur Skeleton." https://www.levins.com/dinosaur.shtml Accessed February 08, 2021.

Levy, J. S., Head, J. W., and Marchant, D. R., 2008: "The Role of Thermal Contraction Crack Polygons in Cold-Desert Fluvial Systems." *Antarctic Science.* **20**, 6: 565–579.

Li, H., Wang, W., Wu, F., Zhan, H., Zhang, G., and Qiu F. 2014. "A New Sand-Wedge–Forming Mechanism in an Extra-Arid Area." *Geomorphology.* 211: 43–51.

Lindgren, A., Hugelius, G., Kuhry, P., Christensen, T. R., and Vandenberghe, J., 2016: "GIS-based Maps and Area Estimates of Northern Hemisphere Permafrost Extent During the Last Glacial Maximum." *Permafrost and Periglacial Processes.* **27**, 1: 6–16.

Little, S., 1946: *The Effects of Forest Fires on the Stand History of New Jersey's Pine Region.* Forest Management Paper No. 2, Northeastern Forest Experiment Station. 43 pp.

Little, S., 1979: "Fire and Plant Succession in the New Jersey Pine Barrens." *In* Forman, R. T. T. (ed.), *Pine Barrens: Ecosystem and Landscape.* New York: Academic Press. pp. 297–314.

Livingstone, I., and Warren, A., 1996: *Aeolian Geomorphology: An Introduction.* Harlow, Essex: Addison Wesley Longman Limited. 211 pp.

Lookingbill, T. R., Brickle, M. C., and Engelhardt, A. M., 2012: *Evaluation of*

Albany Pine Bush, Albany County, New York: For its Merit in Meeting National Significance Criteria as a National Natural Landmark in Representing Sand Dunes as an Example of Eolian Landforms Appalachian Plateau and Ranges Physiographic Regions.* Technical Report, National Park Service National Natural Landmarks Program, 113 pp.

Losco, R. L., Stephens, W., and Helmke, M. F., 2010: "Periglacial Features and Landforms in the Subsurface of Sussex County, Delaware." *Southeastern Geology.* **47**, 2: 85–94.

Lozet, J. M., and Herbillon, A. J., 1971: "Fragipan Soils of Condroz (Belgium): Mineralogical, Chemical and Physical Aspects in Relation with Their Genesis." *Geoderma.* **5**: 325–343.

Lozinski, W. von, 1909: "O Wietrzeniu Mechanicznem Piaskowców w Klimacie Umiarkowanym" (Über die mechanische Verwitterung der Sandsteine im gemäßigten Klima). *Bulletin International de L'Académie des Sciences de Cracovie class des Sciences Mathematique et Naturalles.* **1**: 1–25.

Ludlum, D. M., 1983: *The New Jersey Weather Book.* New Brunswick, NJ: Rutgers University Press. 252 pp.

Lyubtsova, M., 1998: "Small Cryogenic Erosional Relief Forms in the Steppes of Tranbaykalia." *Polar Geography.* **22**, 3: 170–180.

Maarleveld, G. C., 1949: "Over de Erosiedalen van de Veluwe" (On the erosion-valleys of the Veluwe). *Tijdschrift van het Koninklijk Nederlandsch Aardrijkskundig Genootschap.* **66**: 133–142.

Maarleveld, G. C., 1951: "De Asymetrie van de Kleine Dalen op Het Noordelijk Halfrond" (Asymmetry of the small valleys of the northern hemisphere). *Tijdschrift van het Koninklijk Nederlandsch Aardrijkskundig Genootschap.* **68**: 297–312.

Maarleveld, G. C., and van den Toorn, J. C., 1955: "Pseudö-Solle in Noord-Nederland" (Pingo-remnants in the northern Netherlands). *Tijdschrift van het Koninklijk Nederlandsch Aardrijkskundig Genootschap.* **72**, 4: 344–360.

Mackay, J. R., 1979: "Pingos of the Tuktoyaktuk Peninsula Area, Northwest Territories." *Géographie Physique et Quaternaire.* **33**, 1: 3–61.

Mackay, J. R., 1999: "Periglacial Features Developed on the Exposed Lake Bottom of Seven Lakes That Drained Rapidly after 1950, Tuktoyaktuk Peninsula Area, Western Arctic coast, Canada." *Permafrost and Periglacial Processes.* **10**: 39–63.

Magie, D., 2016: *Roman Rule in Asia Minor, Volume 1: To the End of the Third Century After Christ.* Princeton Legacy Library (Book 3800). Princeton, NJ: Princeton University Press. 746 pages.

Makhalanyane, T. P., Valverde, A., Velázquez, D., Gunnigle, E., van Goethem, M. W., Quesada, A., Cowan, D. A., 2015: "Ecology and Biogeochemistry of Cyanobacteria in Soils, Permafrost, Aquatic and Cryptic Polar Habitats." *Biodiversity and Conservation.* **24**: 819–840.

Makkaveyev, A. N., Bronguleev, V. Vad., and Karavaev, V. A., 2015: Short Communication Pleistocene Pingo in the Central Part of the East European Plain. *Permafrost and Periglacial Processes.* **26**, 4: 360–367.

Markewich, H. W., Litwin, R. J., Wysocki, D. A., and Pavich, M. J., 2015: "Synthesis on Quaternary Aeolian Research in the Unglaciated Eastern United States." *Aeolian Research.* **17**: 139–191.

Markewich, H. W., and Markewich, W., 1994: "An Overview of Pleistocene and Holocene Inland Dunes in Georgia and the Carolinas—Morphology, Distribution, Age and Paleoclimate." US Geological Survey Bulletin **2069**. Washington: US Government Printing Office. 31 pp.

Marsh, B., 1987: "Pleistocene Pingo Scars in Pennsylvania." *Geology*. **15**: 945–947.

Marsh, B., 1998: "Wind-Transverse Corrugations in Pleistocene Periglacial Landscapes of Central Pennsylvania." *Quaternary Research*. **49**: 149–156.

Marsh, B., 1999: *Paleoperiglacial Landscapes of Central Pennsylvania. Sixty-second Annual Reunion, Northeast Friends of the Pleistocene 1999 Trip*, Lewisburg, PA: Bucknell University. 69 pp.

Marsh, E., 1979: "The Southern Pine Barrens: an Ethnic Archipelago." In Sinton, J. W. (ed.), *Natural and Cultural Resources of the New Jersey Pine Barrens: Inputs and Research Needs for Planning. Proceedings and Papers of the First Research Conference on the New Jersey Pine Barrens*, Atlantic City, N.J., May 22–23, 1978. Pomona, NJ: Stockton State College. pp. 192–198.

Marsh, E. R., 1985: "A Pleistocene Lake in Central New Jersey." In Talkington, R.W. (ed.), 1985: *Geological Investigations of the Coastal Plain of Southern New Jersey, Part 1: Field Guide. 2nd Annual Meeting of the Geological Association of New Jersey, Geology Program*, Pomona, NJ: Stockton State College. Trenton, NJ: Geological Society of New Jersey. pp. A-14 to A-28.

Marsh, E., Demitroff, M., and Schopp, P., 2019: "The Southern Pine Barrens: an Ethnic Archipelago." *SoJourn: A Journal Devoted to the History, Culture, and Geography of South Jersey*, **3**, 2: 7–25.

Marshall, J. A., Roering, J. J., Rempel, A. W., Shafer, S. L., and Bartlein, P. J., 2021: "Extensive Frost Weathering across Unglaciated North America During the Last Glacial Maximum." *Geophysical Research Letters*. **48**, e2020GL090305.

Martens, J. H. C., 1956: *Industrial Sands of New Jersey*. New Brunswick, NJ: Rutgers University Press. 259 pp.

Martin, E. A., c. 1915: *Dew Ponds: History, Observations and Experiments*. London: T. Werner Laurie. 208 pp.

Martin, G. J., and James, P. E., 1993: *All Possible Worlds: A History of Geographical Ideas*. 3rd edition. New York: John Wiley & Sons. 585 pp.

Martin, R. E., 1998: *One Long Experiment: Scale and Process in Earth Science*. New York: Columbia University Press. 262 pp.

Mason, R. J., 1992: *Contested Lands: Conflict and Compromise in New Jersey's Pine Barrens*. Conflicts in Urban and Regional Development series, volume 6. Philadelphia, PA: Temple University Press. 257 pp.

Mather, J. R., and Sanderson, M., 1996: *The Genius of C. Warren Thornthwaite, Climatologist-Geographer*. Norman and London: University of Oklahoma Press. 226 pp.

Mattocks, R., 1762: Survey and map to 223 acres at a place called The Ponds. West Jersey Surveys, Loose Records 1762, 43884.

McCarty, J., 1757: Survey & map of 170 acres in Gloucester County. West Jersey Loose Records, 1757, 41040.

McConnell, S., 2014: *Witmer Stone: The Fascination of Nature*. Self-published. 424 pp.

McCormick, J., 1955: *A Vegetation Inventory of Two Watersheds in the New Jersey Pine Barrens*. PhD dissertation, Rutgers University, New Brunswick, NJ., 126 pp.

McCormick, J., 1967: *A Study of Significance of the Pine Barrens of New Jersey*. Internal Study for the National Park Service. Philadelphia, PA: The Academy of Natural Sciences of Philadelphia. 100 pp.

McCormick, J., 1970: *The Pine Barrens: a Preliminary Ecological Inventory.* Research Report No. 2. Trenton, NJ: NJ State Museum. 100 pp.

McCormick, J., 1979: "The Vegetation of the New Jersey Pine Barrens." In Forman, R. T. T. (ed.), *Pine Barrens: Ecosystem and Landscape.* New York: Academic Press. pp. 229–243.

McCormick, J., and Andresen, J., 1963: "The Role of *Pinus virginiana* Mill. In The Vegetation of Southern New Jersey." *NJ Nature News.* Audubon Society. **18**: 27–38.

McCormick, J., and Buell, M. F., 1968: "The Plains: Pigmy Forests of the New Jersey Pine Barrens, a Review and Annotated Bibliography." *Bulletin of the New Jersey Academy of Science.* **13**, 1: 20–34.

McDonald, M., 1999: "A Mysterious Blue Swimming Hole—Gouged by a Meteorite?—Has Been Put Back on the Map by Two Old-Timers." *Philadelphia Inquirer.* June 14: B09.

McHorney, R., and Neill, C., 2007: "Alteration of Water Levels in a Massachusetts Coastal Plain Pond Subject to Municipal Ground-Water Withdrawals." *Wetlands.* **27**, 2: 366–380.

McMahon, G., Benjamin, S. P., Clarke, K., Findley, J. E., Fisher, R. N., Graf, W. L., Gundersen, L. C., Jones, J. W., Loveland, T. R., Roth, K. S., Usery, E. L., and Wood, N. J., 2005: *Geography for a Changing World: A Science Strategy for the Geographic Research of the U.S. Geological Survey, 2005–2015.* US Geological Survey Circular 1281, Washington, D.C., 54 pp.

McMahon, W., 1980: *Pine Barrens Legends, Lore and Lies.* Wallingford, PA: Middle Atlantic. 149 pp.

McNeill, J., 1981: "Nomenclatural Problems in *Polygonum*." *Taxon.* **30**, 3: 630–641.

McPhee, J., 1968: *The Pine Barrens.* New York: Farrar, Straus and Giroux. 157 pp.

McQueen, C. B., 1990: *Field Guide to the Peat Mosses of Boreal North America.* University Press of New England. 138 pp.

Mead, M., 1927: "Group Intelligence Tests and Linguistic Disability Among Italian Children." *School and Society.* **25**, 642: 465–468.

Mead, M., 1928: *Coming of Age in Samoa. A Psychological Study of Primitive Youth for Western Civilization.* New York: William Morrow. 170 pp.

Mead, M., 1972: *Blackberry Winter: My Earlier Years.* New York: William Morrow & Co. 305 pp.

Means, J. L., Yuretich, R. F., Crerar, D. A., Kinsman, D. J. J, and Borcsik, M. P., 1981: *Hydrogeochemistry of the New Jersey Pine Barrens.* Bulletin 76, NJ Geological Survey. 107 pp.

Melin, B. E., and Graves, R. C., 1971: "The Water Beetles of Miller Blue Hole, Sandusky County, Ohio (Insecta: Coleoptera)." *The Ohio Journal of Science.* **71**, 2: 73–77.

Melnikov, V. P., and Spesivtsev, V. I., edited by Konishchev V. N. 2000. *Cryogenic Formations in the Earth's Lithosphere.* (graphic version). Novosibirsk: Scientific Publishing Center of the UIGGM, Siberian Branch of the RAS, Publishing House, Siberian Branch of the RAS. 343 pp. In Russian (pp. 1–172) and English (pp. 173–343).

Meredith, D., 2002: "Hazards in the Bog—Real and Imagined." *Geographical Review.* **92**, 3: 319–332.

Meredith, H., 1731a: "Remainder of the Account of Cape Fear Begun in Our :ast." *Pennsylvania Gazette.* May 13, 1731, Accessible Archives item #700.

Meredith, H., 1731b: "Account of Cape Fear." *Pennsylvania Gazette.* May 6, 1731, Accessible Archives item #691.

Merritts, D., Marshall, J., Walter, R.,

Demitroff, M., Hertzler, N., Ruck, J., Blair, A., Huot, S., Alter, S., Corbett, L. B., Caffee, M. W., Bierman, P. R., and Eppes, M., In review: "Evidence of LGM Permafrost and Thaw, Eastern US: Relict Thermal Contraction Cracking Polygons, Sand Wedges, and Gelifluction Lobes." Submitted to *Geology*, August, 2021.

Merritts, D., Walter, R., Blair, A., Demitroff, M., Potter, N., Alter, S., Markey, E., Guillorn, S., Gigliotti, S., and Studnisky, C., 2015: "LiDAR, Orthoimagery, and Field Analysis of Periglacial Landforms and Their Cold Climate Signature, Unglaciated Pennsylvania and Maryland." Geological Society of America. *Abstracts with Programs* **47**, 7: 831.

Merritts, D. J., and Rhanis, M. A., 2022: "Pleistocene Periglacial Processes and Landforms, Mid-Atlantic Region, Eastern United States." *Annual Review of Earth and Planetary Sciences.* **50**: 541–592.

Meyer, B. K., Vance, R. K., Bishop, G. A., and Deocampo, D. M., 2014; "Origin and Dynamics of Nearshore Wetlands: Central Georgia Bight, USA." *Wetlands.* **35**, 2: 247–261.

Meyer, J. L., and Youngs, Y., 2019: "Historical Landscape Change in Yellowstone National Park: Demonstrating the Value of Intensive Field Observation and Repeat Photography." *Geographical Review.* **108**: 387–409.

Michel, F. A., 1994: "Changes in Hydrogeologic Regimes in Permafrost Regions Due to Climatic Change," *Permafrost and Periglacial Processes.* **5**: 191–195.

Mickle, I., 1845: *Reminiscences of Old Gloucester or Incidents in the History of the Counties of Gloucester, Atlantic and Camden, New Jersey.* Philadelphia: Townsend Ward. 170 pp. (1968 reprint, Gloucester County Historical Society).

Middleton, B., 2002: "Nonequillibrium Dynamics of Sedge Meadows Grazed by Cattle in Southern Wisconsin." *Plant Ecology.* **161**: 89–110.

Middleton, B., 2004: "Cattle Grazing and Its Long-Term Effects on Sedge Meadows." Fact Sheet 2004-3027, US Geological Survey, 4 pp.

Mihalasky, M. J., and Del Sontro, T. S., 2003: "Spung Map: Great Egg Harbor Watershed Region, Southern New Jersey." *In* Hozik, M. J., Mihalasky, M. J. (eds.), *Field Guide and Proceedings, 20th Annual Meeting of the Geological Association of New Jersey, October 10–11, 2003*. Trenton, NJ: Geological Society of New Jersey. pp. 107–110.

Mikhailov, V. M., 1998: "Thermal Regime of Northern Rivers and Its Relation to the Development of Taliks." *Permafrost, Engineering Geology and Mining Ecology, International Conference on Arctic Margins 1994.* Conference Proceedings, Magadan, Russia, pp. 332–337. https://www.boem.gov/94ICAM332-337/ Accessed April 08, 2018.

Mikhailov, V. M., 2016: "Flood-Plain Taliks in the Valleys of Meandering Rivers in Northeastern Russia." *Earth's Cryosphere.* **20**, 2: 37–44.

Miller, E., 1749: *A Map of the Honorable Thomas Penn's and Richard Penn's Land on the Prince Morris River, 1749.* Millville Historical Society, Millville, New Jersey, 1 sheet.

Miller, G., 1781: Survey of 40 acres in Gloucester County, West Jersey Surveys, Book R, Folio 91.

Miller, J. S., and Hoel, L. A., 2002: "The 'Smart Growth' Debate: Best Practices for Urban Transportation Planning." *Socio-Economic Planning Sciences.* **36**: 1–24.

Minard, J. P., 1966: "Sandblasted Blocks on a Hill in the Coastal Plain of New Jersey." *Geological Survey Research 1966, Chapter B.* Professional Paper

550-B, pp. B87–B-90.

Minard, J. P., and Rhodehamel, E. C., 1969: "Quaternary Geology of Part of Northern New Jersey and the Trenton Area." In Subitzky, S. (ed.), *Geology of Selected Areas in New Jersey and Eastern Pennsylvania and Guidebook of Excursions*. New Brunswick, NJ: Rutgers University Press. pp. 279–313.

Mol, J., 1997: "Fluvial Response to Weichselian Climate Changes in the Niederlausitz (Germany)." *Journal of Quaternary Science*. **12**, 1: 43–60.

Mol, J., Vandenberghe, J., and Kasse, C., 2000: "River Response to Variations of Periglacial Climate in Mid-Latitude Europe." *Geomorphology*. **33**: 131–148.

Moonsammy, R. Z., Cohen, D. S., and Williams, L. E., 1987: *Pinelands Folklife*. New Brunswick, NJ: Rutgers University Press. 234 pp.

Moor, E., 1823: *Suffolk Words and Phrases; or, an Attempt to Collect the Lingual Localisms of that County*. Woodbridge, UK: Printed by J. Loder for R. Hunter. 561 pp.

Moore, C. R., Brooks, M. J., Mallinson, D.J., Parham, P. R., Ivester, A. H., and Feathers, J. K., 2016: "The Quaternary Evolution of Herndon Bay, a Carolina Bay on the Coastal Plain of North Carolina (USA): Implications for Paleoclimate and Oriented Lake Genesis." *Southeastern Geology*. **51**: 145–171

Moore, H., 1943: *An Old Jersey Furnace: A Study*. Baltimore, MD: Newth-Morris Publishing. 15 pp.

Moore, K., 2002a: "Drought Hastening Drying of Pinelands." *Asbury Park Press*. B1, B4

Moore, K., 2002b: "How Dry We Are." *New Jersey Monthly*. September: 88–130.

Morse, P. D., and Wolfe, S. A., 2017: "Long-Term River Icing Dynamics in Discontinuous Permafrost, Subarctic Canadian Shield." *Permafrost and Periglacial Processes*. **28**: 580–586.

Mounier, R. A., 1972: "The Blue Hole site." *Bulletin of the Archeological Society of New Jersey*. **29**: 14–27.

Mounier, R. A., 2003: *Looking Beneath the Surface: the Story of Archaeology in New Jersey*. New Brunswick, NJ: Rutgers University Press. 261 pp.

Mounier, R. A., 2008: *The Aboriginal Exploitation of Cuesta Quartzite in Southern New Jersey*. PhD dissertation, Memorial University, St. John's, NFL., 435 pp.

Mounier, R. A., and Wilson, B., 2006: "Tar Kiln, Gloucester Furnace: an Archeological Tar Kiln near Pomona, Galloway Township, Atlantic County, New Jersey." http://www.njpinelandsanddownjersey.com/open/index.php?module=documents&JAS_DocumentManager_op=viewDocument&JAS_Document_id=63. Accessed December 16, 2006.

Mountford, K., 2002: *Closed Sea: From the Manasquan to the Mullica, a History of Barnegat Bay*. Harvey Cedars, NJ: Down the Shore Publishing. 207 pp.

Mowszowicz, J., and Olaczek, R., 1961: "Flora Naczyniowa Rezerwatu 'Niebieskie Zródla'" (Vascular flora of the "Blue Springs" reserve). *Lodzkie Towarzystwo Naukowe, Societas Scientiarum Lodziensis*. **73**, 3, 3: 1–40.

Mulhouse, J. M., De Steven, D., Lide, R. F., and Sharitz, R. R., 2005: "Effects of Dominant Species on Vegetation Change in Carolina Bay Wetlands Following a Multi-Year Drought." *Journal of the Torrey Botanical Society*. **132**, 3: 411–420.

Murton, J. B., and Ballantyne, C. K., 2017: Chapter 5: "Periglacial and Permafrost Ground Models for Great Britain." In Griffiths, J. S., and Martin, C. J., *Engineering Geology and Geomorphology of Glaciated and Periglaciated Terrains* –

Engineering Group Working Party Report. Geological Society, London, Engineering Geology Special Publications. **28**: 501–597.

Murton, J. B., and Belshaw, R. K., 2011: "A Conceptual Model of Valley Incision, Planation and Terrace Formation During Cold and Arid Permafrost Conditions of Pleistocene Southern England." *Quaternary Research*. **75**: 385–394.

Murton, J. B., and French, H. M., 1993: "Thermokarst Involutions, Summer Island, Pleistocene Mackenzie Delta, Western Canadian Arctic." *Permafrost and Periglacial Processes*. **4**: 217–229.

Murton, J. B., and Lautridou, Jean-Pierre, 2003: "Recent Advances in the Understanding of Quaternary Periglacial Features of the English Channel Coastlands." *Journal of Quaternary Science*. **18**, 3–4: 301–307.

Murton, J. B., Worsley, P., and Gozdzik, J., 2000: "Sand Veins and Wedges in Cold Eolian Environments." *Quaternary Science Reviews*. **19**: 899–922.

Mycielska-Dowgiallo, E., 1977: "Channel Pattern Changes During the Last Glaciation and Holocene, in the Northern Part of the Sandomierz Basin and the Middle Part of the Vistula Valley, Poland." *In* Gregory, K.J. (ed.), *River Channel Changes*. Chichester, UK: John Wiley & Sons. pp. 75–87.

Myers, G. A., McGreevy, P., Carney, G. O., and Kenny, J., 2003: "Cultural Geography." *In* Gaile, G. L., Willmott, C. J. (eds.), *Geography in America at the Dawn of the 21st Century*. Oxford: Oxford University Press. pp. 81–96.

Mylroie, J. E., Carew, J. L., and Moore, A. I., 1995: "Blue Holes: Definition and Genesis." *Carbonates and Evaporites*. **10**, 2: 225–233.

National Wetlands Working Group, 1997: *The Canadian Wetland Classification System*, 2nd Edition. Warner, B. G. and Rubec, C. D. A. (eds.), Wetlands Research Centre, University of Waterloo, Waterloo, ON, Canada. 68 pp.

National Park Service. 2018: "Bethel AME Church. Aboard the Underground Railroad: A National Register Travel Itinerary." https://www.nps.gov/nr/travel/underground/nj3.htm. (Accessed August 08, 2018).

Nationaler Geopark, 2022: "Nationale GeoParks." http://www.nationaler-geopark.de/startseite.html (Accessed July 14, 2022).

Navoy, A. S., 2004: "Pinelands Hydrology." *New Jersey Flows*. New Jersey Water Resources Research Institute Special Pinelands Edition. **5**, 2: 4–5.

Nelson, S., 1781: Map of Amos Strettels 900, 1680 and part of 4390 acres of land in Broad Neck: Salem County Historical Society, Salem, NJ, 1 sheet.

Nelson, F. E., Hinkel, K. M., Shiklomanov, N. I., Meuller, G. R., Miller, L. L., and Walker, D. A., 1998: "Active-Layer Thickness in North Central Alaska: Systematic Sampling, Scale, and Spatial Autocorrelation." *Journal of Geophysical Research*. **103**, D22: 28,963–28,973.

Nelson, F. E., Hinkel, K. M., and Outcalt, S. I., 1992: "Palsa-Scale Frost Mounds." *In* Dixon, J. C., and Abrahams, A. D., *Periglacial Geomorphology: Proceedings of the 22nd Annual Binghamton Symposium in Geomorphology*. Chichester, UK: John Wiley & Sons. pp. 305–325.

Nelson, F. E., Shiklomanov, N. I., Meuller, G. R., Hinkel, K. M., Walker, D. A., and Bockheim, J. G., 1997: "Estimating Active-Layer Thickness Over a Large Region: Kuparuk River Basin, Alaska, U.S.A." *Arctic and Alpine Research*. **29**, 4: 367–378.

Neuendorf, K. K. E., Mehl, Jr., J. P., and Jackson, J. A., 2005: *Glossary of*

Geology. 5th Edition. Alexandria, VA: American Geological Institute. 779 pp.

New Jersey Board of Agriculture, 1893: *Twentieth Annual Report of the State Board of Agriculture: 1892–1893*. Document No. 25, Trenton, NJ, 528 pp. + appendix.

New Jersey Historical Records Survey Project, 1940: *Transcriptions of Early Church Records of New Jersey: Colporteur Reports to the American Tract Society 1841–1846*. Historical Records Survey, Works Project Administration, Newark, NJ, 124 pp.

New Jersey Secretary of State, 1734: This is a true and perfect enveytorey [inventory] of all the goods and chatles of James Steelma [n] late deceased. Estate Files, Inventory of James Stellman, Will No. 182H, Eggharbour, January 4th day, 1734/1735, Trenton, NJ: New Jersey State Archives.

New Jersey Secretary of State, 1736: A true and perfect inventory of the moveable estate of Andrew Steelman late of Eggharbour. Estate Files, Inventory of Andrew Stellman, Will No. 641H, Eggharbour, February the 3rd, 1736/37. Trenton, NJ: New Jersey State Archives.

Newell, W. L., 2005: "Evidence of Cold Climate Slope Processes from the New Jersey Coastal Plain: Debris Flow Stratigraphy at Haines Corner, Camden County, New Jersey." US Geological Survey, Open-File Report 2005-1296. July 2005. Version 1.0. Unpaginated. http://pubs.usgs.gov/of/2005/1296/ Accessed August 09, 2018.

Newell, W. L., and DeJong, B. D., 2011: "Cold-Climate Slope Deposits and Landscape Modifications of the Mid-Atlantic Coastal Plain, Eastern USA." *In* Martini, I. P., French, H. M., Pérez, A. A. (eds.), *Ice-Marginal and Periglacial Processes and Sediments*. Geological Society, London, Special Publications, **354**, pp. 259–276.

Newell, W. L., Powers, D. S., Owens, J. P., Stanford, S. D., and Stone, B. D., 2000: *Surficial Geologic Map of Central and Southern New Jersey*. US Geological Survey, Miscellaneous Investigations Series, Map I-2540-D.

Newell, W. L., and Wyckoff, J. S., 1992: "Paleohydrology of Four Watersheds in the New Jersey Coastal Plain." *In* Gohn, G. S. (ed.), *Proceedings of the 1988 U.S. Geological Survey Workshop on the Geology and Geohydrology of the Atlantic Coastal Plain*. US Geological Survey, Circular 1059. pp. 23–28.

Newell, W. L., Wyckoff, J. S., Owens, J. P., and Farnsworth, J., 1989: *Southeast Friends of the Pleistocene, 2nd Annual Field Conference: Cenozoic Geology and Geomorphology of Southern New Jersey Coastal Plain, November 11–13, 1988*. US Geological Survey, Open File Report 89-159.

Nichols, P. H., 1953: *Periglacial Ventifacts in New Jersey*. M.S. thesis, Rutgers University, New Brunswick, NJ, 46 pp.

Nichols, R. L., 1969: "Geomorphology of Inglefield Land, North Greenland." *Meddelelser om Grønland*. **188**, 1: 1–109.

Nickles, A., c.1958: *An Historic Trip up the Great Egg Harbor River*. M.S. thesis, Glassboro State College, Glassboro, NJ, 182 pp.

NJDEP (New Jersey Department of Environmental Protection), c. 1931, Airplane atlas sheets. Bureau of Tidelands, Trenton, NJ. aerial photomosaics. Scale 1" = 1000', 261 Sheets.

NJDEP (New Jersey Department of Environmental Protection), 2003a: *Status of the Water Supply of Southeastern New Jersey*. draft, September 2003. Land Use Management, Division of Watershed Management, Trenton, NJ. 77 pp.

NJDEP (New Jersey Department of

Environmental Protection), 2003b: Status of the Water Supply of Southeastern New Jersey: Executive Summary. draft, September 2003. Land Use Management, Division of Watershed Management, Trenton, NJ. 18 pp.

NJDEP (New Jersey Department of Environmental Protection), 2019: Surface water springs of New Jersey. Bureau of GIS. https://gisdata-njdep.opendata.arcgis.com/datasets/surface-water-springs-of-new-jersey?page=8 (Accessed February 10, 2021).

Norton, W., 1984: *Historical Analysis in Geography*. London: Longman Group. 231 pp.

Olsson, R. K., 2001: "Peter Wolfe." *The Redbeds: The Annual Newsletter of the Department of Geological Sciences, Rutgers, The State University of NJ*. **5**: 3.

Oxford University Press. Various dates. Various entries. In *Oxford English Dictionary*.

Otvos, E. G., 1973: *Guidebook: Geology of the Mississippi-Alabama Coastal Area and Nearshore Zone*. The New Orleans Geological Society, New Orleans, Louisiana. 67 pp.

Otvos, E. G., 1995: "Geomorphic Development and Paleoenvironments of Late Pleistocene Sand Hills, Southeastern Louisiana: Discussion and Reply." *Southeastern Geology*. **35**, 4: 211–223.

Otvos, E. G., 2015: "The Last Interglacial Stage: Definitions and Marine Highstand, North America and Eurasia." *Quaternary International*. **383**: 158–173.

Owens, J. P., Hess, M. M., Denny, C. S., and Dwornik, E. J., 1983: "Postdepositional Alteration of Surface and Near-surface Minerals in Selected Coastal Plain Formations of the Middle Atlantic States." US Geological Survey Professional Paper 1067-F, 45 pp.

Paepe, R., and Pissart, A., 1969: "Periglacial Structures in the Late-Pleistocene Stratigraphy of Belgium." *Biuletyn Peryglacjalny*. **20**: 321–333.

Palmer, M., 2005: *The Effects of Microtopography on Environmental Conditions, Plant Performance, and Plant Community Structure in Fens of the New Jersey Pinelands*. Ph.D. dissertation. New Brunswick, NJ: Rutgers, The State University of New Jersey. 159 pp.

Parker, G. G., Higgins, C. G., and Wood, W. W., 1990: "Piping and Pseudokarst in Drylands." In Higgins, C. G., and Coates, D. R. (eds.), *Groundwater Geomorphology; the Role of Subsurface Water in Earth-Surface Processes and Landforms*. Geological Society of America Special Paper 252. pp. 77–110.

Parsekian, A., and Slater, L., 2011: Report as of FY2010 for 2010NJ230B: "Hydrogeophysical Investigation of Subsurface Controls on Persistent Canopy Gaps in the New Jersey Pinelands." Water Resources Research Act, Program Details for Project ID 2010NJ230B, 7 pp.

Parsons, B., 1767: Survey of 38 acres in Great Egg Harbor Township, Gloucester County, NJ. West Jersey Surveys, Book N, Folio 338.

Pastore, C. L., Green, M. B., Bain, D. J., Munoz-Hernandez, A., Vörösmarty, C. J., Arrigo, J., Brandt, S., Duncan, J. M., Greco, F., Kim, H., Kumar, S., Lally, M., Parolari, A. J., Pellerin, B., Salant, N., Schlosser, A., and Zalzal, K., 2010: "Tapping Environmental History to Recreate America's Colonial Hydrology." *Environmental Science & Technology*. **2010**, 44: 8798–8803.

Paul, J. D., 2014: "The Relationship Between Spring Discharge, Drainage, and Periglacial Geomorphology of the Frome valley, Central Cotswolds, UK." *Proceedings of the Geologists'*

Association. **125**, 2: 182–194.

Paul, S. P., 1822: Survey of 1470.04 acres in Galloway Township, Gloucester County, NJ. West Jersey Loose Records, 58063.

Pearson, M., 1994: *A Place That's Known*. Jackson, MS: University Press of Mississippi. 256 p.

Peltier, L. C., 1949a: "There Appeared to Be a General Agreement." In Assembled notes from the "Friends" Pensauken field trip, May. pp. 5–6. Notes on file at the NJ Geological Survey, Trenton, NJ.

Peltier, L. C., 1949b: *Pleistocene terraces of the Susquehanna River, Pennsylvania*. PA Geological Survey Fourth Series, Bulletin G 23. Harrisburg, PA: Topographic and Geologic Survey. 158 pp.

Penn, R., 1750: Thomas Penn & Richard Penn's 6896 acres of Land at Morris River on their warn't for 48:000 acres on the fifth Dividend. West Jersey Loose Records, NJ State Archives # 59873.

Pennypacker, J. L., 1936: *Verse and Prose of James Lane Pennypacker*. Haddonfield, NJ: Historical Society of Haddonfield. 242 pp.

Pepper, A., 1971: *The Glass Gaffers of New Jersey, and Their Creations from 1739 to the Present*. New York: Scribner. 330 pp.

Perennou, C., Guelmami, A., Paganini, M., Philipson, P., Poulin, B., Strauch, A., Tottrup, C., Truckenbrodt, J., and Geijzendorffer, I. R., 2018: "Mapping Mediterranean Wetlands with Remote Sensing: A Good-Looking Map Is Not Always a Good Map." *Advances in Ecological Research*, **2018**: 243–277.

Perkins, S., 2007: "Not-So-Perma Frost: Warming Climate Is Taking Its Toll in Subterranean Ice." *Science News*. **171**, 20: 154.

Petway, P. J., 1961: "History of Negroes in Vineland." *The Vineland Historical Magazine*. 61–64.

Péwé, T. L., 1973: "Ice-Wedge Casts and Past Permafrost Distribution in North America." *Geoforum*. **15**: 15–26.

Péwé, T. L., 1983: "The Periglacial Environment in North America During Wisconsin time." In Porter, S. C. (ed.), *Late Quaternary Environments of the United States, Volume 1. The Late Pleistocene*. Minneapolis, MN: University of Minnesota Press. pp. 157–189.

Philhower, C. A., 1931: "South Jersey Indians on the Bay, the Cape and the Coast." *Proceedings of the New Jersey Historical Society*. **16**: 1–21.

Pierce, A. D., 1957: *Iron in the Pines: The Story of New Jersey's Ghost Towns and Bog Iron*. New Brunswick, NJ: Rutgers University Press. 244 pp.

Pierce, A. D., 1960: *Smugglers' Woods: Jaunts and Journeys in Colonial and Revolutionary New Jersey*. New Brunswick, NJ: Rutgers University Press. 322 pp.

Pierce, A. D., 1964. *Family Empire in Jersey Iron: the Richards Enterprises in the Pine Barrens [by] Arthur D. Pierce*. New Brunswick, NJ: Rutgers University Press. 286 pp.

Pierzchalko, L., 1954: P. E. Wolfe: "Periglacial Frost-Thaw Basins in New Jersey." *Sprawozdania z Literatury, Biuletyn Peryglacjalny*. **1**: 98–99.

Pinchot, G., 1899: "The Plains." In Geological Survey of New Jersey. *Annual Report of the State Geologist for the Year 1889: Report on Forests*. Trenton, NJ: MacCrellish & Quigley. pp. 125–130.

Pinelands Commission, 2015a: "The Comprehensive Management Plan." Pinelands Comprehensive Management Plan (CMP). https://www.nj.gov/pinelands/cmp/. Accessed August 09, 2018.

Pinelands Commission, 2015b: "Kirkwood-Cohansey Project."

https://www.nj.gov/pinelands/science/complete/kc/ Accessed August 14, 2018.

Pinkava, D. J., 1963: "Vascular Flora of the Miller Blue Hole and Stream, Sandusky County, Ohio." *The Ohio Journal of Science.* **63**, 3: 113–127.

Pissart, A., 1965: "Les Pingos des Hautes Fagnes: Les Problèms de Leur Genese" (Pingos of Hautes Fagnes: problems of their genesis). *Annales de la Société Géographique de Belgique.* **88**, 5–6: B277–B289.

Pissart, A., 1970: "Les Phénomènes Physiques Essentiels Liés au Gel, Les Structures Périglaciaires qui en Résultent et Leur Signification Climatique" (Basic physical phenomena related to freezing, the resultant periglacial structures, and their climactic significance). *Annales de la Société Géologique de Belgique.* **93**: 7–49.

Pissart, A., 1974: "Les Viviers des Hautes Fagnes Sont des Traces de Buttes Periglaciaires mais S'agissait-Il Réellement de Pingos" (The fishponds of High Fagnes are traces of periglacial hillocks. But are they the result of pingos)? *Annales de la Société Géologique de Belgique.* **97**: 359–381.

Pissart, A., 2000: "Remnants of Lithalsas of the Hautes Fagnes, Belgium: a Summary of Present-Day Knowledge." *Permafrost and Periglacial Processes.* **11**: 327–355.

Pissart, A., 2003: "The Remnants of Younger Dryas Lithalsas on the Hautes Fagnes Plateau in Belgium and Elsewhere in the World." *Geomorphology.* **52**: 5–38.

Platt, I. H., 1889: "The Pine Belt of New Jersey: a Region of Sandy Soil and Pine Forests." *Transactions of the American Climatological Association.* June 1889. 9 pp.

Polhamas, A., 2015: "N.J.'s 'Blue Holes' Are a Beautiful, Deadly Temptation for Swimmers, Experts Say." NJ.com http://www.nj.com/gloucester-county/index.ssf/2015/07/south_jersey_blue_holes_are_a_beautiful_deadly_tem.html accessed March 31, 2018.

Post, M. B., 2014: "Kirkwood-Cohansey Aquifer's Health Critical to Pinelands Ecosystems." *The Press of Atlantic City.* April 13.

Poster, L. S., Rhoads, A. F., and Block, T. A., 2013: "Vascular Flora and Community Assemblages of Delhaas Woods, a Coastal Plain Forest in Bucks County, Pennsylvania." *The Journal of the Torrey Botanical Society.* **140**: 101–124.

Powell, G., 1958: "Blue Hole"—A problem in geology. In the Gloucester County Historical Society Collection, in a file titled "Blue Hole." 6 pp.

Powley, V. R., 1969: *Soil Survey of Salem County, New Jersey.* Washington, DC: US Department of Agriculture. 86 pp.

Prince, H. C., 1961: "Some Reflections on the Origin of Hollows in Norfolk Compared with Those in the Paris Region." *Revue de Géomorphologie Dynamique.* **12**: 110–117.

Prince, H. C., 1962: "Pits and Ponds in Norfolk." *Erdkunde.* **16**: 10–31.

Prince, H. C., 1964: "The Origin of Pits and Depressions in Norfolk." *Geography.* **49**: 15–32.

Prince, H. C., 1979: "Comments." In Prince, H. C., Jennings, J. N., Sperling, C. H. B., Goudie, A. S., Stoddart, D. R., and Poole, G. G., Discussion: "Marl Pits or Dolines of the Dorset Chalklands?" *Transactions of the Institute of British Geographers.* **4**, 1: 116–124.

Prowell, G. R., 1886: *The History of Camden County, New Jersey.* Philadelphia, PA: L. J. Richards. 769 pp.

Pugsley, A. J., 1939: *Dewponds in Fable and Fact.* London: Country Life. 62 pp.

Purvis, W. J., 1903: "Traditions of Malaga Road." *Valley Ventura.* **VIII**, 380–383,

Sept. 18 & 25, October 2 & 9.
Purvis, W. J., 1917: "The Old Cohansey Road." *The Vineland Historical Magazine.* **3**: 27–29.
Purvis, W. J., 1920: "One Hundred Years of History of the Old Malaga Road." *The Vineland Historical Magazine.* **5**: 134–138.
Pye, K., 1995: "The Nature, Origin and Accumulation of Loess." *Quaternary Science Review.* **14**: 653–667.
Radis, R., 1987: "New Jersey Plants: a Fen is a Bog is a Swale is a Spong . . ." *NJ Audubon,* **12**, 4: 18–19.
Rakowska, B., 1996: "Diatom Communities occurring in Niebieskie Zródla near Tomaszów Mazowiecki, Central Poland (1963–1990)." *Fragmenta Floristica et Geobotanica* **41**, 2: 639–655.
Rasmussen, W. C., 1953: "Periglacial Frost-Thaw Basins in New Jersey: a Discussion." *Journal of Geology.* **61**: 473–474.
Rasmussen, W. C., 1958: *Geology and Hydrology of the "Bays" and Basins of Delaware.* Ph.D. dissertation, Bryn Mawr, PA: Bryn Mawr College. 206 pp.
Reimer, C. W., 1959: "The Diatom Genus *Neidium.* I. New Species, New Records, and Taxonomic Revisions." *Proceedings of the Academy of Natural Sciences of Philadelphia.* **111**: 1–36.
Reimer, C. W., Henderson, M. V., and Mahoney, R. K., 1991: "Contributions of Charles S. Boyer (1856–1928) to the Knowledge of Diatoms (Bacillariophyceae): Biographical Notes, Literature and Taxonomic Summary, with Type Designations." *Proceedings of the Academy of Natural Sciences of Philadelphia.* **143**: 161–172.
Repelewska-Pekalowa, J., 1991: "Periglacial Morphogenesis of the Coastal Plains of Recherche Fiord (Spitsbergen)." *Wyprawy Geograficzne na Spitsbergen, UMCS, Lublin.* **1991**: 45–56.
Repelewska-Pekalowa, J., 1996: "Development of Relief Affected by Contemporary Geomorphological Processes in NW part of Wedel Jarlsberg Land (Bellsund, Spitsbergen-Svalbard)." *Biuletyn Peryglacjalny.* **35**: 153–181.
Rhodehamel, E. C., 1970: *A Hydrologic Analysis of the New Jersey Pine Barrens Region.* Water Resources Circular. No. 22. State of New Jersey, Division of Water Policy and Supply. 35 pp.
Rhodehamel, E. C., 1973: *Geology and Water Resources of the Wharton Tract and the Mullica River Basin in Southern New Jersey.* Special Report No. 36, Division of Water Resources. Trenton, NJ: State of New Jersey Department of Environmental Protection. 58 pp.
Rhodehamel, E. C., 1979a: "Geology of the Pine Barrens of New Jersey." In Forman, R.T.T. (ed.), *Pine Barrens: Ecosystem and Landscape.* New York: Academic Press. pp. 39–60.
Rhodehamel, E. C., 1979b: "Hydrology of the New Jersey Pine Barrens." In Forman, R. T. T. (ed.), *Pine Barrens: Ecosystem and Landscape.* New York: Academic Press. pp. 147–167.
Rhoads, B. L., 2004: "Whither Physical Geography?" *Annals of the American Geographers.* **94**, 4: 748–755.
Ribeiro, W. G., 1996a: " 'The Blue Hole' of Winslow Township." *South Jersey Magazine.* **25**, 2: 6–8.
Ribeiro, W. G., 1996b: "The BLUE HOLE revisited." *South Jersey Magazine.* **25**, 3: 42–43.
Richards, H. G., and Judson, S., 1965: "The Atlantic Coastal Plain and the Appalachian Highlands in the Quaternary." In Wright, H. E., Jr., Frey, D. G. (eds.), *The Quaternary of the United States.* INQUA Review vol. for the VII Cong., Princeton, NJ: Princeton University Press. pp. 129–136.
Richards, H. G., and Rhodehamel, E., 1965: New Jersey Coastal Plain field

trip (August 17). *In* Schultz, C. B., Smith, H. T. U. (eds.), *Guidebook for Field Conference B-1, Central Atlantic Coastal Plain, Seventh INQUA Congress, Boulder, Colorado.* pp. 10–13.

Richardson, C. J. (ed.), 1981: *Pocosin Wetlands: An Integrated Analysis of Coastal Plain Freshwater Bogs in North Carolina.* Stroudsburg, PA: Hutchinson Ross Publishing. 364 pp.

Ricklefs, R. E., 1990: *Ecology.* New York, NY: Freeman. 896 pp.

Risley, D. L?, 1896: *Estelle and Milmay: The Growing Colony in southern New Jersey, Views of Pioneer Life.* South Jersey Land & Transportation Co. 25 pp. (reprinted in 2020 by the South Jersey Culture & History Center, Stockton University, Galloway, NJ).

Roberson, J. A., 2018: "My 'Inside-the-Beltway' Crystal Ball Is Broken." *Journal of the American Water Works Association.* **110**: 45–49.

Roberts, M., Reiss M., and Monger, G., 1993: *Biology: Principles and Processes.* Surry, UK: Nelson. 852 pp.

Roman, C. T., and Good, R. E., 1983: *Wetlands of the New Jersey Pinelands: Values, Functions, and Impacts.* Division of Pinelands Research, Center for Coastal and Environmental Studies. New Brunswick, NJ: Rutgers University. 82 pp.

Roman, C. T., Zampella, R. A., and Jaworski, A. Z., 1985: "Wetland Boundaries in the New Jersey Pinelands: Ecological Relationships and Delineation." *Water Resources Bulletin.* **21**, 6: 1005–1012.

Rose, T. F., Price, T. T., and Woolman, H. C., 1878: *Historical and Biographical Atlas of the New Jersey Coast.* Philadelphia, PA: Woolman & Rose. 372 pp. facsimile edition by the Ocean County Historical Association. (1985).

Ross, T. E., 2000: *Carolina Bays: An Annotated and Comprehensive Bibliography 1844–2000.* Southern Pines, NC: Carolinas Press. 113 pp.

Rotnicka, J., and Rotnicki, K., 1988: "The Problem of the Hydrological Interpretation of Palaeochannel Pattern." *In* Lang, G., and Schlüchter, C. (eds.), *Lake, Mire and River Environments During the Last 15000 years: Proceedings of the INQUA/IGCP 158 Meeting on the Palaeohydrological Changes During the Last 15000 Years, Bern, June 1985.* Rotterdam, NETH: A. A. Balkema. pp. 205–224.

Royal Colony of New Jersey, 1765: An Act to enable the Honourable Charles Read, Esq; to erect a Dam over Batstow Creek, and also to enable John Estell to erect a Dam over Atsion River, 20 June 1765

Russell, E. W. B., and Krug, E., 1980: "Landscape Features and Bog Iron Ore Deposits of the New Jersey Pine Barrens." *In* Manspeizer, W. (ed.)., *Field Studies of New Jersey Geology and Guide to Field Trips: 52nd Annual Meeting of the New York State Geological Association.* Geology Department, Newark College of Arts & Sciences. Newark, NJ: Rutgers University. pp. 146–157.

Russell, E. W. B., and Stanford, S. D., 2000: "Late-Glacial Environmental Changes South of the Wisconsinan Terminal Moraine in the Eastern United States." *Quaternary Research.* **53**: 105–113.

Sagan, C., Veverka, J., Fox, P., Dubisch, R., Lederberg, J., Levinthal, E., Quam, L., Tucker, R., Pollack, J. B., and Smith, B. A., 1972: "Variable Features on Mars: Preliminary Mariner 9 Television Results." *Icarus.* **17**: 346–372.

Salem County Tercentenary Committee, 1964: *Fenwick's Colony.* Salem, NJ: Salem County Tercentenary Committee. 191 pp.

Salisbury, R. D. 1893. "Part I. Surface Geology. Report of Progress." *In Geological Survey of New Jersey.*

Annual Report of the State Geologist of New Jersey for 1892. Trenton, NJ: John L. Murphy. pp. 34–166.

Salisbury, R. D., 1898: The *Physical Geography of New Jersey*. N.J. Geological Survey Final Report, v. 4, 170 p.

Salisbury, R. D., and Knapp, G. N., 1917: *The Quaternary Formations of Southern New Jersey*. Volume VII of the final report series of the State Geologist, Department of Conservation and Development. Trenton, NJ: Division of Geology and Waters. 218 pp.

Sanford, H. D., Sawyer, S. C., Kocher, S. D., Hiers, J. K., and Cross, M. 2017: "Linking Knowledge to Action: the Role of Boundary Spanners in Translating Ecology." *Frontiers in Ecology and the Environment*. **15**, 10: 560–568.

Sauer, C. O., 1941: "Foreword to Historical Geography." *Annals of the Association of American Geographers*. **31**: 1–24.

Sauer, C. O., 1956: "The Education of a Geographer." *Annals of the Association of American Geographers*. **46**: 287–299.

Savage, H., 1982: *The Mysterious Carolina Bays*. Columbia, SC: University of South Carolina Press. 121 pp.

Scagnelli, R., and Moore, G., 2009: 20 August: Milmay area, Atlantic, Cumberland, and Cape May Counties, New Jersey. Joint trip with the Torrey Botanical Society. *Bartonia*: Journal of the Philadelphia Botanical Club. **64**: 62–63

Sceurman, M., and Moran, M., 2004: *Weird N.J.: Your Travel Guide to New Jersey's Local Legends and Best Kept Secrets*. New York: Barnes & Noble. 372 pp.

Schaetzl, R. J., and Anderson, S., 2005: *Soils: Genesis and Geomorphology*. Cambridge: Cambridge University Press. 817 pp.

Schirmer, W. (ed.), 1999: "Dunes and Fossil Soils." *GeoArcaeoRhein*. **3**: 1–190.

Schmitthenner, S., 1926: "Die Entstehung der Dellen und Ihre Morphologische Bedeutung" (the emergence of the dellen and their morphological meaning) *Zeitschrift für Geomorphologie*. **1**: 3–28.

Schokker, J., 2003: "Patterns and Processes in a Pleistocene Fluvio-Aeolian Environment: Roer Valley Graben, South-Eastern Netherlands." *Nederlandse Geografische Studies* **314**. Koninklijk Nederlands Aardrijkskundig Genootschap/ Faculteit Ruimtelijke Wetenschappen, Unviversiteit Utrecht, Utrecht. 142 pp.

Schuyler, A. E., 1990: Element stewardship abstract for *Narthecium americanum*. Stewardship Abstract No. 010, State of New Jersey, Department of Environmental Protection and Energy, Trenton, NJ. https://www.njparksandforests.org/natural/heritage/textfiles/nartham.txt (Accessed February 14, 2021).

Schwabe, S., and Herbert, R. A., 2004: "Black Holes of the Bahamas: What They Are and Why They Are Black." *Quaternary International*. **121**: 3–11.

Schwans, E., Meng, T. M., Prudhomme, K., and Morgan, M. L., 2015: "On the Origin of the Crestone Crater: Low-Latitude Periglacial Features in San Luis Valley, Colorado." EP53A-1002, American Geophysical Union Fall Meeting, San Francisco, Abstract.

Schlyter, P., 1994: "Paleo-Periglacial Ventifact Formation by Suspended Silt or Snow–Site Studies in South Sweden." *Geografiska Annaler: Series A, Physical Geography*. **76**, 3: 187–201.

Scientific American Correspondence, 1896: "The Sandstorm in New Jersey." *Scientific American*. March 21, 1896: 183

Seer, F. K., and Schrautzer, J., 2014: "Status, Future Prospects, and Management Recommendations for Alkaline Fens in an Agricultural Landscape: A Comprehensive Survey." *Journal for Nature Conservation.* **22**, 4: 358–368.

Seppälä, M., 1971: "Evolution of Eolian Relief of the Kaamasjoki-Kiellajoki River Basin in Finnish Lapland." *Publicationes Instituti Geographici Universitatis Turkuensis.* **54**: 1–88. (republished *Fennia.* **104**).

Seppälä, M., 2004: *Wind as a Geomorphic Agent in Cold Climates: Studies in Polar Research.* Cambridge University Press: Cambridge. 358 pp.

Sheinkman, V. S., 2012: "Specificity in Development of Glaciation in the High Mountains of Siberia—an Ecological Aspect." *Tomsk State University Journal of Biology.* **2**: 210–231.

Sheinkman, V. S., 2016: "Quaternary Glaciation in North-Western Siberia— New Evidence and Interpretation." *Quaternary International.* **420**: 15–23.

Shimwell, D. W., 1971: *The Description and Classification of Vegetation.* Seattle, WA: University of Washington Press. 322 pp.

Sidorchuk, A. Y., Panin, A. V., Borisova, O. K., Elias, S. A., and Syviski, J. P., 2000: "Channel Morphology and River Flow in the Northern Russian Plain in the Late Glacial and Holocene." *International Journal of Earth Sciences.* **89**: 541–549.

Siman-Tov, S., Stock, G. M., Brodsky, E. E., and White, J. C., 2017: "The Coating Layer of Glacial Polish." *Geology*, 11: 987–990.

Sinton, J. W., (ed.). 1978: *Natural and Cultural Resources of the New Jersey Pine Barrens: Inputs and Research Needs for Planning. Proceedings and Papers of the First Research Conference on the New Jersey Pine Barrens, Atlantic City, N.J., May 22–23, 1978.* Pomona, NJ: Stockton State College. 365 pp.

Sim, R. J., and Weiss, H. B., 1955: *Charcoal-Burning in New Jersey from Earl Times to the Present.* Trenton, NJ: New Jersey Agricultural Society. 62 pp.

Skinner, A., and Schrabisch, M., 1913: *A Preliminary Report of the Archaeological Survey of the State of New Jersey. Geological Survey of New Jersey*, Bulletin 9. Trenton, NJ: MacCrellish & Quigley. 94 pp.

Small, J. A., 1952: "The Pine Barrens of New Jersey: Introduction." *Bartonia, Journal of the Philadelphia Botanical Club.* **26**: 19.

Small, R. J., Clark, M. J., and Lewin, J., 1970: "The Periglacial Rock-Stream at Clatford Bottom, Marlborough Downs, Wiltshire." *Proceedings of the Geologists' Association.* **81**: 87–98.

Smith, D., 2012: *Succession Dynamics of Pine Barrens Riverside Savannas.* MS thesis, New Brunswick, NJ: Rutgers

Smith, E., 1773: Survey of 131.4 acres in Maurice River Township, Cumberland County, NJ. West Jersey Surveys Book P, folio 332.

Smith, H. T. U., 1949: "Physical Effects of Pleistocene Climatic Changes in Nonglaciated Areas: Eolian Phenomena, Frost Action, and Stream Terracing." *Bulletin of the Geological Society of America.* **60**: 1485–1516.

Smith, J., 2020: "Salty Livestock." https://www.smithjam.com/salty-livestock/ (Accessed February 10, 2021).

Smith, J. B., 1910: *The Insects of New Jersey.* Annual Report of the New Jersey State Museum, 1909. MacCrellish and Quigley, Trenton, NJ. 888 pp.

Smith, S., 1765: *The History of the Colony of Nova-Caearia, or New-Jersey: Containing, an Account of its First Settlement, Progressive Improvements, the Original and Present Constitution, and other Events, to the Year 1721. With some Particulars Since; and a Short View of its Present State.* Burlington,

NJ: James Parker. 573 pp.

Smolová, I., 2002: "Geomorphologic Analysis of the Javorí Hory Mountains Relief." *Geographica.* **37**: 83–90.

Sneddon, L. A., Metzler, K. J., and Anderson, M., 1995: "A Classification and Description of Natural Community Alliances and Selected Community Elements of the Delaware Estuary." *In* Dove, L. E., and Nyman, R. M. (eds.), *Living Resources of the Delaware Estuary.* The Delaware Estuary Program. pp. A-3 to A-90.

Snyder, D. B. V., and Vivian, V. E., 1981: *Rare and Endangered Vascular Plant Species in New Jersey.* Newton Corner, MA: U.S. Fish & Wildlife Service. 89 pp.

Sokolov, I. A., Alekseyev, V. P., Markov, M. L., and Kolotayev, V. I., 1989: "Research into Icings and Icing Processes in the USSR: Major Results and Prospects." *Polar Geography and Geology.* **13**, 4: 233–251.

Somber, S., Snook, I., and Hoffman, J. L., 2013: "Using the Low Flow Margin Method to Assess Water Availability in New Jersey's Water-Table-Aquifer Systems." NJ Geological and Water Survey, Technical Memorandum 13-3, NJ Department of Environmental Protection, Trenton, NJ, 70 pp.

Southgate, (Russell) E. W. B., 2000: "Vegetation and Fire History of a Pine Barren Riverside Savanna: Above Buck Run, Oswego River Drainage Basin, Pine Barrens, New Jersey." *In:* Walz, K. S., Stanford, S., Boyle, J., Southgate, (Russell) E. W. B., *Pine Barren Riverside Savannas of New Jersey.* New Jersey Department of Environmental Protection, Division of Parks and Forestry, Office of Natural Lands Management, Natural Heritage Program, Trenton, NJ. 169 pp. + appendices.

Sparks, B. W., Williams, R. B. G., and Bell, F. G., 1972: "Presumed Ground-Ice Depressions in East Anglia." *Proceedings of the Royal Society of London, Series A, Mathematical and Physical Sciences.* **327**, 1570: 329–343.

Staff of the New Jersey Pinelands Commission, 1984: "The Ecological Implications of Exporting Water from the Cohansey Aquifer." Proceedings of a technical advisory committee meeting on the issue of pumping Cohansey water to meet the water supply needs of the metropolitan Camden area. New Lisbon, NJ: New Jersey Pinelands Commission. 19 pp. + draft review comments.

Stanford, S. D., 1997: "Road Log and Description of Field Stops, Saturday May 31, 1997." *In* Stanford S. D., and Witte R. W. (leaders), *Pliocene-Quaternary Geology of Northern New Jersey: Guidebook for the 60th Annual Reunion of the Northeastern Friends of the Pleistocene.* NJ Geological Survey. pp. 6-1 to 6-31.

Stanford, S. D., 2003: "Late Miocene to Holocene Geology of the New Jersey Coastal Plain." *In* Hozik, M. J., Mihalasky, M. J. (eds.), *Field Guide and Proceedings, 20th Annual Meeting of the Geological Association of New Jersey, October 10–11, 2003.* Trenton, NJ: Geological Association of New Jersey. pp. 21–49.

Stanford, S. D., 2005: "The Geologic History of New Jersey's Landscape." *Unearthing New Jersey.* **1**, 2: 1–4.

Stanford, S. D., 2017: *Geology of the Oswego Lake Quadrangle, Burlington and Ocean Counties, New Jersey.* Open File Map 118, New Jersey Geological Survey, scale: 1:24,000, map.

Stanford, S. D., Stone, B. D., Ridge, J. C., Witte, R. W., Pardi, P. R., and Reimer, G. E., 2021: "Chronology of Laurentide Glaciation in New Jersey and the New York City Area, United States." *Quaternary Research.* **99**: 142–167.

Starkel, L., 1995: "The Place of the Vistula River Valley in the Late Vistulian—

Early Holocene Evolution of the European Valleys." In Frenzel, B. (ed.), 1995: *European River Activity and Climatic Change During the Lateglacial and Early Holocene*. Special Issue: ESF Project "European Palaeoclimate and Man" 9. *Paläoklimaforschung/ Palaeoclimate Research*. **14**: 1–226.

Starkel, L., Gregory, K. J., and Thornes, J. B. (eds.), 1991: *Temperate Palaeohydrology: Fluvial Processes in the Temperate Zone During the Last 15,000 Years*. Chichester, UK: John Wiley & Sons. 548 pp.

Steelman, G., 1741/1742: Survey of 100 acres at a place called Lochs of the Swamp and Great Neck, Great Egg Harbor Township, NJ. West Jersey Surveys, Bulls Book, Folio 115.

Steelman, J. F., 2003: "Location of the Mill of Elisha Smith (1706-1755)." *Atlantic County Historical Society Yearbook*. **14**: 256–259.

Steelman, R. B., 1965: *Cumberland's Hallowed Heritage*. Bridgeton, NJ: Evening News. 55 pp.

Steinhausen, J., 1936: *Archäologische Siedlungskunde des Trierer Landes. Herausgegeben vom Rheinischen Landesmuseum Trier*. Trier Druck und Verlag der Paulinus-Druckerei GmbH 1936. 614 pp.

Stevenson, J. R., 1890: "John Hart 'the Signer.'" *The New York Genealogical and Biographical Record*. **21**: 36–39.

Stewart, D. J., 1876: *Combination Atlas Map of Cumberland County, New Jersey*. Philadelphia, PA: Stewart. 45 pp. (reprinted by the Greenwich, NJ: Cumberland County Historical Society. 2004).

Stewart, F. H., 1918: *Stewart's Genealogical and Historical Miscellany No 2*. Woodbury, NJ: Gloucester County Historical Society. 72 pp. (index added in 1971 reprint).

Stewart, F. H., 1932: *Indians of Southern New Jersey*. Woodbury, NJ: Gloucester County Historical Society. 93 pp.

Still, J., 1877: *Early Recollections and Life of Dr. James Still*. New Brunswick, NJ: Rutgers University Press. 278 pp. (Reprinted 1973).

Stokes, W. L., 1968: "Multiple Parallel-Truncated Bedding Planes—A Feature of Wind-Deposited Sandstone Formations." *Journal of Sedimentary Petrology*. **38**, 2: 510–515.

Stolt, M. H., and Rabenhorst, M. C., 1987a: "Carolina Bays on the Eastern Shore of Maryland: I. Soil Characterization and Classification." *Soil Science Society of America Journal*. **51**: 394–398.

Stolt, M. H., and Rabenhorst, M. C., 1987b: "Carolina Bays on the Eastern Shore of Maryland: II. Distribution and Origin." *Soil Science Society of America Journal*. **51**: 399–405.

Stone, J. R., and Ashley, G. M., 1992: Ice-Wedge Casts, Pingo Scars, and the Drainage of Glacial Lake Hitchcock." In Robinson, P., and Brady, J. B. (eds.), *Guidebook for Field Trips in the Connecticut Valley Region of Massachusetts and Adjacent States*, v. 2 (84th Annual Meeting of the New England Intercollegiate Geological Conference), Amherst Massachusetts. University of Massachusetts, Geology Department, Contribution 66. pp. 305–331.

Stone, W., 1911: "The Plants of Southern New Jersey with Especial Reference to the Flora of the Pine Barrens and the Geographic Distribution of the Species." In Morse, S. R., curator, *Annual Report of the New Jersey State Museum Including a Report of the Plants of Southern New Jersey, with Especial Reference to the Flora of the Pine Barrens: 1910*. Trenton, NJ: MacCrellish & Quigley. pp. 23–828.

Stone, W., 1937: *Bird Studies at Old Cape May: An Ornithology of Coastal New Jersey*. 2 Volumes. New York: Dover Publications. (1965 reprint).

Stott, P., 2005: Convention concerning the protection of the world cultural and natural heritage. UNTS 15511, World Heritage Convention, United Nations Educational, Scientific and Cultural Organization, Adapted by the General Conference at its seventeenth session Paris, 16 November 1972. 17 pp.

Strahler, A. N., 1950: "Equilibrium Theory of Erosional Slopes Approached by Frequency Distribution Analysis." *American Journal of Science.* **248**, 10: 673–696.

Strahler, A. N., 1952: "Hypsometric (area-altitude) Analysis of Erosional Topography." *Geological Society of America Bulletin.* **63**, 11: 1117–1142.

Stutz, B., 2017: "In a Rare U.S. Preserve, Water Pressures Mount as Development Closes In." *YaleEnvironment360*. Published at the Yale School of the Environment. Conservation. https://e360.yale.edu/features/in-a-rare-u-s-preserve-water-pressures-mount-as-development-closes-in (Accessed February 12, 2021).

Sugden, D., and Hall, A., 2020: "Antarctic Blue-Ice Moraines: Analogue for Northern Hemisphere ice sheets?" *Quaternary Science Reviews.* **249**, 106620.

Sui, D., and DeLyser, D., 2012: "Crossing the Qualitative–Quantitative Chasm I: Hybrid Geographies, the Spatial Turn, and Volunteered Geographic Information (VGI)." *Progress in Human Geography.* **36**, 1: 111–124.

Sun, G., Callahan, T. J., Pyzoha, J. E., and Trettin, C. C., 2006: "Modeling the Climatic and Subsurface Stratigraphy Controls on the Hydrology of a Carolina Bay Wetland in South Carolina, USA." *Wetlands.* **26**, 2: 567–580.

Suther, B. E., Leigh, D. S., Brook, G. A., and Yang, L., 2018: "Mega-Meander Paleochannels of the Southeastern Atlantic Coastal Plain, USA." *Palaeogeography, Palaeoclimatology, Palaeoecology.* **511**: 52–79.

Svenska Akademiens Ordbok, 1952: *Panna*. Volume 19, p. 127. https://www.saob.se/artikel/?unik=P_0001-0387.tkl2&pz=3 Accessed March 30, 2018.

Swain, R., 1792: *Journal of Rev. Richard Swain While on the Salem Circuit, N.J.* (transcribed and edited by Steelman, R. B., Old First United Methodist Church, West Long Branch, NJ, January 12, 1977).

Swezey, C. S., 2020: "Quaternary Eolian Dunes and Sand Sheets in Inland Locations of the Atlantic Coastal Plain Province, USA." *In* Lancaster, N., and Hesp, P. (eds.), *Inland Dunes of North America*, Dunes of the World, Springer Nature Switzerland, pp.11–63, https://doi.org/10.1007/978-3-030-40498-7_2.

Swezey, C. S., 2018: "Carolina Bays of the U.S. Atlantic Coastal Plain are Relict Thermokarst Lakes that Formed Episodically During the Last Glaciation." *Geological Society of America Abstracts with Programs.* **50**, 3.

Swezey, C. S., Schultz, A. P., González, W. A., Bernhardt, C. E., Doar, W. R. III, Garrity, C. P., Mahand, S. A., and McGeehina, J. P., 2013: "Quaternary Eolian Dunes in the Savannah River Valley, Jasper County, South Carolina, USA." *Quaternary Research.* **80**, 2: 250–264.

Swezey, C. S., Fitzwater, B. A., Whittecar, G. R., Mahan, S. A., Garrity, C. P., Gonzalez, W. B. A., and Dobbs, K. M., 2016: "The Carolina Sandhills: Quaternary Eolian Sand Sheets and Dunes Along the Updip Margin of the Atlantic Coastal Plain Province, Southeastern United States." *Quaternary Research.* **86**: 271–286.

Szczypek, T., and Waga, J. M. (eds.), 1996:

Wspolczesne oraz Kopalne Zjawiska i Formy Eoliczne Wybrane Zagadnienia (Present-Day and Fossil Aeolian Phenomena and Landforms: Selected Problems)*. Wybrane zagadnienia. Wydzial Nauk o Ziemi Uniwersytetu Śląskiego, Park Krajobrazowy "Cysterskie Kompozycje Krajobrazowe Run Wielkich," Stowarzyszenie Geomorfologów Polskich, Sosnowiec, POL. 169 pp.

Szumanski, A., and Starkel, L., 1990: "Reconstruction of Both Palaeohydrological Elements and Palaeoclimate Based on Environmental Features." *In* Starkel, L. (ed.), *Evolution of the Vistula River Valley During the Last 15000 Years, Part III.* Geographical Studies Special Issue No. 5. Wroclaw, POL: Ossolineum, The Publishing House of the Polish Academy of Sciences. pp. 154–163.

Taylor, W. G. (ed.), 1879: "Recently While Some Men Were Surveying in Atlantic County . . ." *May's Landing Record.* **2**, 40: 3, 07/19/1879.

Taylor, W. G. (ed.), 1880a: "Sheriff's Sale. By Virtue of a Writ . . ." *May's Landing Record.* **3**, 29: 3, 05/01/1880.

Taylor, W. G. (ed.), 1880b: "Does Our Friend in Attempting to Write the History of Our Own Beautiful Inland Town . . ." *May's Landing Record.* **3**, 38: 3, 07/03/1880.

Tedrow, J. C. F., 1963: *New Jersey Soils*. Circular 601, College of Agriculture and Environmental Science. New Brunswick, NJ: Rutgers University. 20 pp.

Tedrow, J. C. F., 1973: "Polar Soil Classification and the Periglacial Problem." *Biuletyn Peryglacjalny.* **22**: 285–294.

Tedrow, J. C. F., 1979: "Development of Pine Barrens Soils." *In* Forman, R. T. T. (ed.), *Pine Barrens: Ecosystem and Landscape*. New York: Academic Press. pp. 61–79.

Tedrow, J. C. F., 1986: *Soils of New Jersey*. Malabar, FL: Robert E. Krieger. 480 pp.

Tedrow, J. C. F., Bruggeman, P. F., and Walton, G. F., 1968: *Soils of Prince Patrick Island*. Research Paper 44. The Arctic Institute of North America. 82 pp.

Tedrow, J. C. F., and MacClintock, P., 1953: "Loess in New Jersey Soil Materials." *Soil Science.* **75**, 1: 19–29.

Thatcher, W. A. (Engineer), 1919: Tract no. 41 farm lots at Richland Garden Farms Extension no. 2 near Mays Landing on West Jersey and Seashore Electric Line owned by Gilbert & Callaghan, 703 Walnut Street, Philadelphia, Penna. 600 feet = 1 inch. Atlantic County Clerk's Office, File #786, filed February 20, 1924.

Thiagarajan, N., and Lee, Cin-Ty A., 2004: "Trace-Element Evidence for the Origin of Desert Varnish by Direct Aqueous Atmospheric Deposition." *Earth and Planetary Science Letters.* **224**: 131–141.

Thomas, G., 1698: *An Historical and Geographical Account of the Province and Country of Pennsylvania: and of West-New-Jersey in America*. London.

Thomas, J. B., 1940: "The Wild Cattle of the Beaches." *In* Weygandt, C., *Down Jersey: Folks and Their Jobs, Pine Barrens, Salt Marsh and Sea Islands*. New York: D. Appleton-Century. pp. 300–304.

Thompson, D. G., 1928: *Groundwater Supplies of the Atlantic City Region*. Bulletin 30, Reports of the Department of Conservation and Development. Trenton, NJ: Division of Waters. 138 pp.

Thornthwaite, C. W., 1940: "The Relation of Geography to Human Ecology." *Ecological Monographs.* **10**, 3: 343–348.

Tiner; R. W., Jr., 1985: *Wetlands of New Jersey*. U.S. Fish and Wildlife Service, National Wetlands Inventory, Newton Comer, MA. 117 pp.

Tomaso, M. S., Bello, C. A., Dillian, C., and Demitroff, M., 2008: Geoarchaeological analysis of Prehistoric site 28-BU-718, Township of Evesham. Burlington County, New Jersey (Abstract). Eastern States Archaeological Federation, 75th Annual Meeting, November 6-9, 2008, p. 28.

Tomlin, C., 1932: *Tomlin Genealogy*. Eakins, Palmer, & Harrar: Philadelphia. 91 pp.

Toms, E., Mason, P. J., and Ghail, R. C., 2014: "Drift-Filled Hollows in Battersea: Investigation of the Structure and Geology Along the Route of the Northern Line Extension, London." *Quarterly Journal of Engineering Geology and Hydrogeology*. **49**: 147-153.

Toms, H. S., 1927: Ancient ponds near Cissbury. *Sussex Magazine*. **1**: 404-407.

Toms, H. S., 1934a: Ancient ponds near Patcham. *Sussex Magazine*. **8**: 486-90.

Toms, H. S., 1934b: "Ancient Ponds Near Falmer: Broad Shackles and Buckland Bank." *Sussex Magazine*. **8**: 546-550.

Townsend, T., Corson, S., and Commis, A. B., 1873: Peck's Beach: Somers Survey, 488 acres, Dower Line of Anna R. Somers, September 25, 1873. pp 97-98.

Trela, J. J., 1984: *Soil Formation on Tertiary Landsurfaces of the New Jersey Coastal Plain*. Ph.D. dissertation, New Brunswick, NJ: Rutgers. 620 pp.

Tremblay, L. P., 1961: "Wind Striations in Northern Alberta and Saskatchewan, Canada." *Geological Society of America Bulletin*. **72**: 1561-1564.

Troll, C., 1957: "Tiefenerosion, Seitenerosion und Akkumulation der Flüsse im Fluvioglazialen und Periglazialen Bereich" (Downward erosion, lateral erosion, and the river accumulations in glacial and periglacial zones). In Louis, H. (ed.), *Geomorphologische Studien*. Fritz Machatschek zum 80. Geburtstag gewidmet von Schülern, Freuden, Verehrern und dem Verlag. 213-226.

Troll, C., 1962: "Sölle" and "Mardelles." *Erdkunde*. **16**: 31-34.

Troll, C., 1973: "Rasenabschälung (turf exfoliation) ala Periglaziales Phänomen der Subpolaren Zonen und der Hochgebirge" (Turf exfoliation as a periglacial phenomenon of subpolar zones and alpine regions). *Zeitschrift für Geomorphologie*, Neue Folge, Supplementband. **17**: 1-32.

Troetschel, H. Jr., 1950: "John Clayton Gifford: An Appreciation." *Tequesta*. **10**: 35-47.

Trusty, E. M., 1997: *The Underground Railroad—Ties that Bound Unveiled, A History of the Underground Railroad in Southern New Jersey from 1770 to 1861*. Philadelphia, PA: Amed Literary. 422 pp.

Tsoar, H., and Møller, J. T., 1986: "The Role of Vegetation in the Formation of Linear Sand Dunes." In Nickling, W. H. (ed.), *Aeolian Geomorphology*. Boston: Allen and Unwin. pp. 75-95.

Turner II, B. L., 2005: "Geography's Profile in Public Debate 'Inside the Beltway' and the National Academies." *The Professional Geographer*. **57**, 3: 2005.

van Amerom, H. W. J. (ed.), 1990: "Tundra Rivers of the Last Glacial: Sedimentation and Geomorphological Processes During the Middle Pleniglacial in Twente, Eastern Netherlands/Dynamics of Vegetation and Environment During the Middle Pleniglacial in the Dinkel Valley (The Netherlands)/The Dinkel Valley in the Middle Pleniglacial: Dynamics of a Tundra River System." *Mededelingen Rijks Geologische Dienst*. **44**, 3: 1-138.

van Balen, R. T., Kasse, C., and De Moor, J., 2008: "Impact of Groundwater Flow on Meandering; Example from the Geul River, The Netherlands." *Earth Surface Processes and Landforms*. **33**:

2010–2028.

van der Valk, A. G., 1981: "Succession in Wetlands: a Gleasonian Approach." *Ecological Society of America.* **63**, 3: 688–696.

van Everdingen, R. V., 1998: *Multi-Language Glossary of Permafrost and Related Ground-Ice Terms* (2005 revision). Calgary, AB: University of Calgary. 90 pp. +illustrations & references.

van Huissteden, J. K., and Vandenberghe, J., 1988: "Changing Fluvial Style of Periglacial Lowland Rivers During the Weichselian Pleniglacial in the Eastern Netherlands." *Zeitschrift für Geomorphologie.* **32**, suppl. 71: 131–146.

van Kerkhoff, L., and Lebel, L., 2006: "Linking Knowledge and Action for Sustainable Development." *Annual Review of Environment and Resources.* **31**: 445–477.

van Rensselaer, S., and Boyer, C. S., 1926: "New Jersey Glass Houses." *In* van Rensselaer, S., *Early American Bottles and Flasks*. Revised Edition. Petersborough, NH: Transcript Printing. pp. 115–155.

van Vliet-Lanoë, B., 1983: *Etudes Cryopédologiques au Sud du Kongsfjord, Svalbard.* Publication interne du Centre de Géomorphologie du CNRS, Caen. 39 pp.

van Vliet-Lanoë, B., 1996: "Relations Entre la Contraction Thermique des Sols en Europe du Nord-Ouest et la Dynamique de l'Inlandsis Weichsélien" (Soil thermal contraction in NW Europe related to the dynamics of the Weichselian ice sheet). *Comptes Rendus de l'Académie des Sciences.* Série II. Fascicule a. **322**: 461–468.

van Vliet-Lanoë, B., 2005: *La Planète des Glaces. Histoire et Environnements de Notre ère Glaciaire* (The Ice Planet: History and Environments of the Glacial Era). Vuibert: Paris. 470 pp.

van Vliet-Lanoë, B., 2006: "Deformations in the Active Layer Related with Ice/Soil Wedge Growth and Decay in Present Day Arctic." *Paleoclimatic implications. Annales de la Société Géologique du Nord.* **13**, 2: 81–95.

van Vliet-Lanoë, B., Brulhet, J., Combes, P., Duvail, C., Ego, F., Baize, S., and Cojan, I., 2017: "Quaternary Thermokarst and Thermal Erosion Features in Northern France: Origin and Palaeoenvironments.' *Boreas.* **46**, 3: 442–461.

van Vliet-Lanoë, B., and Langohr, R., 1981: "Correlation Between Fragipans and Permafrost with Special Reference to Silty Weichselian Deposits in Belgium and Northern France." *Catena.* **8**: 137–154.

van Wely, F. P. H. P., 1973: *Cassell's English-Dutch, Dutch-English Dictionary.* New York: Funk and Wagnalls. 1354 pp.

Vandenberghe, J., 1995: "Timescales, Climate and River Development." *Quaternary Science Reviews.* **14**: 631–638.

Vandenberghe, J., 2001: "A Typology of Pleistocene Cold-Based Rivers." *Quaternary International.* **79**: 111–121.

Vandenberghe, J., 2002: "The Relation Between Climate and River Processes, Landforms and Deposits During the Quaternary." *Quaternary International.* **91**: 17–23.

Vandenberghe, J., 2008: "The Fluvial Cycle at Cold-Warm-Cold Transitions in Lowland Regions: A refinement of theory." *Geomorphology.* **98**: 275–284.

Vandenberghe, J., Bohncke, S., Lammers, W., and Zilverberg, L., 1987: "Geomorphology and Palaeoecology of the Mark Valley (Southern Netherlands): Geomorphological Valley Development During the Weichselian and Holocene." *Boreas.* **16**: 55–67.

Vandenberghe, J., French, H. M., Gorbunov, A., Marchenko, S., Velichko,

A. A., Jin, H., Cui, Z., Zhang, T., and Wan, X., 2014: "The Last Permafrost Maximum (LPM) Map of the Northern Hemisphere: Permafrost Extent and Mean Annual Air Temperatures, 25–17 ka BP." *Boreas*. **43**, 3: 652–666.

Vandenberghe, J., and Sidorchuk, A., 2020: "Large Palaeomeanders in Europe: Distribution, Formation Process, Age, Environments and Significance." *In*: Herget, J., and Fontana, A. (eds.)., *Palaeohydrology. Traces, Tracks and Trails of Extreme Events*. Geography of the Physical Environment. Cham: Springer. https://doi.org/10.1007/978-3-030-23315-0_9.

Vandenberghe, J, and van den Broek, P., 1982: "Weichselian Convolution Phenomena and Processes in Fine Sediments." *Boreas*. **11**: 299–315.

Vandenberghe, J., Wang, X., and Vandenberghe, D., 2016: "Very Large Cryoturbation Structures of Last Permafrost Maximum Age at the Foot of the Qilian Mountains (NE Tibet Plateau, China)." *Permafrost and Periglacial Processes*. **27**, 1: 138–143.

Vandenberghe, J., and Woo, M.-K., 2002: "Modern and Ancient Periglacial River Types." *Progress in Physical Geography*. **26**: 479–506.

Várkonyi, P. L., Laity, J. E., Domokos, G. D. 2016. "Quantitative Modeling of Facet Development in Ventifacts by Sand Abrasion." *Aeolian Research*. **20**: 25–33.

Veillette, J. J., and Thomas, R. D., 1979: "Icings and Seepage in Frozen Glaciofluvial Deposits, District of Keewatin, N.W.T." *Canadian Geotechnical Journal*. **16**: 789–798.

Velichko, A. A., and Zelikson, E. M., 2005: "Landscape, Climate and Mammoth Food Resources in the East European Plain During the Late Paleolithic Epoch." *Quaternary International*. **126–128**: 137–151.

Vermeule, C. C., 1894: *Report on Water-Supply, Water-Power, the Flow of Streams and Attendant Phenomena*. Volume III, of the Final Report to the State Geologist. Trenton, NJ: Geological Survey of New Jersey. 352 pp.

Wacker, P. O., 1979: "Human Exploitation of the New Jersey Pine Barrens Before 1900." *In* Forman, R. T. T. (ed.), *Pine Barrens: Ecosystem and Landscape*. New York: Academic Press. pp. 3–23.

Wah, J. S., Wagner, D. P., and Lowery, D.L., 2018: "Loess in the Mid-Atlantic Region, USA." *Quaternary Research*. **89**: 786–796

Waksman, S. A., 1942: *The Peats of New Jersey and Their Utilization*. Bulletin No.55, Department of Conservation and Development, State of New Jersey: Trenton. 155 pp.

Waksman, S. A., Schulhoff, H., Hickman, C. A., Cordon, T. C., and Stevens, S. C., 1943: *The Peats of New Jersey and Their Utilization*. Bulletin No.55-Part B, Department of Conservation and Development. Trenton: State of New Jersey. 278 pp.

Waldykowski, P., and Krzemien, K., 2013: "The Role of Road and Footpath Networks in Shaping the Relief of Middle Mountains on the Example of the Gorce Mountains (Poland)." *Zeitschrift für Geomorphologie*, Neue Folge. **57**, 4: 429–470.

Walsh, D., 2003: "Getting to the Bottom of the Myths of the Blue Hole in the Pine Barrens." *Courier-Post*. February 20: 1B.

Walters, J. C., 1975: *Origin and Paleoclimatic Significance of Fossil Periglacial Phenomena in Central and Northern New Jersey*. Ph.D. dissertation, New Brunswick, NJ: Rutgers University. 137 pp.

Walters, J. C., 1978: "Polygonal Patterned Ground in Central New Jersey. *Quaternary Research*. **10**: 42–54.

Walters, J. C., 1983: "Sorted Patterned Ground in Ponds and Lakes of the High Valley/Tangle Lakes Region, Central Alaska." In National Research Council. *Proceedings, Fourth International Conference on Permafrost, Volume One*. National Academy Press: Washington, DC. pp. 1350–1355.

Walz, K. S., Stanford, S., Boyle, J., and Southgate, W. F. R., 2006: *Pine Barrens Riverside Savannas of New Jersey*. New Jersey Department of Environmental Protection, Division of Parks and Forestry, Office of Lands Management, Natural Heritage Program, Trenton, NJ. 169 pp.

Warner, A. G., 1869: *Sketches, Incidents and History: Vineland and the Vinelanders*. Vineland, NJ: Crocker Steam Job Printer (reprinted in *The Vineland Historical Magazine*, **57**: 18–71).

Washburn, A. L., 1979: *Geocryology: A Survey of Periglacial Processes and Environments*. London: Edward Arnold. 406 pp.

Watson, W., 1812: *A Map of the State of New Jersey. To His Excellency, Joseph Bloomfield, Governor, the Council and Assembly of the State of New Jersey: This Map Is Respectfully Inscribed*. Philadelphia, PA: W. Harrison, September 25, Scale: 4 miles to 1 inch.

Watts, W. A., 1979: "Late Quaternary Vegetation of Central Appalachia and the New Jersey Coastal Plain." *Ecological Monographs*. **49**: 427–469.

Watts, W. A., 1983: "Vegetational History of the Eastern United States, 25,000 to 10,000 Years Ago." In Porter, S. C. (ed.), *Late Quaternary Environments of the United States, Volume 1. The Late Pleistocene*. Minneapolis, MN: University of Minnesota Press. pp. 294–310.

Weiss, H. B., and Kemble H. R., 1962: *They Took to the Waters: The Forgotten Mineral Spring Resorts of New Jersey and Nearby Pennsylvania and Delaware*. Trenton, NJ: The Past Times Press. 232 pp.

Weiss, H. B., and Weiss, G. M., 1965: *Some Early Industries of New Jersey*. Trenton, NJ: New Jersey Agricultural Society. 70 pp.

Weiss, H. B., and Weiss, G. M., 1968: *The Early Sawmills of New Jersey*. Trenton, NJ: New Jersey Agricultural Society. 98 pp.

Weygandt, C., 1940: *Down Jersey: Folks and Their Jobs, Pine Barrens, Salt Marsh and Sea Islands*. New York: D. Appleton-Century. 352 pp.

Wherry, E. T., 1957: "Goose Pond Saved." *Bartonia, Journal of the Philadelphia Botanical Club*. **29**: 2.

White, F., 1785: *The Philadelphia Directory*. Philadelphia, PA: Young, Stewart, and McCulloch.

White, J. J., 1870: *Cranberry Culture*. New York: Orange Judd Co. 126 pp.

White, J. J., 1885: *Cranberry Culture*. New and enlarged edition. New York: Orange Judd Co. 131 pp.

White, J. J., 2014: "Cranberry Culture. A First-Hand Account of the Development of the J. J. White Company." *Whitesbog Preservation Trust Newsletter*. **4th Quarter**: 1–15. (reprint adaptation of a J. J. White account dated October 15, 1914).

Whitehead, W. A., 1875: *East Jersey Under the Proprietary Governments: a Narrative of Events Connected with the Settlement and Progress of the Province, until the Surrender of the Government to the Crown in 1703. Drawn Principally from Original Sources. With An Appendix Containing "The Model of the Government of East Jersey, in America" Reprinted from the Original Edition of 1685*. 2nd ed. Newark, NJ: Martin R. Dennis. 486 pp.

Whittaker, R. H., 1979: "Vegetational Relationships of the Pine Barrens." In

Forman, R. T. T. (ed.), *Pine Barrens: Ecosystem and Landscape.* New York: Academic Press. pp. 315–331.

Wichansky, P. S., Weaver, C.P., Steyaert, L. T., and Walko, R. L., 2006: "Evaluating the Effects of Historical Land Cover Change on Summertime Weather and Climate in New Jersey." In Maher, N. M. (ed.), *New Jersey's Environments: Past, Present and Future.* New Brunswick, NJ: Rutgers University Press. pp. 128–163.

Widmer, K., 1964: *The Geology and Geography of New Jersey.* Volume 19, The New Jersey Historical Series. Princeton, NJ: D. Van Nostrand. 193 pp.

Wilkinson, K. N., 2003: "Colluvial Deposits in Dry Valleys of Southern England as Proxy Indicators of Paleoenvironmental and Land-Use Change." *Geoarchaeology*—An International Journal. **18**, 7: 725–755.

Williams, P. J., and Smith, M.W., 1989: *The Frozen Earth: Fundamentals of Geocryology.* Cambridge: Cambridge University Press. 306 pp.

Williams, R. B. G., 1973: "Frost and the Works of Man." *Antiquity.* **47**: 19–31

Williams, R. G. B., 1987: "Periglacial Phenomena in the South Downs." In Sieveking, de G., Hart, M. B. (eds.), *The Scientific Study of Flint and Chert.* Proceedings of the 4th International Flint Symposium. Cambridge: Cambridge University Press. 161–167.

Willis, L. L. T., Balliet L. D., and Fish, M. R. M. (eds.), 1915: *Early History of Atlantic County.* Somers Point, New Jersey: Atlantic County Historical Society. 179 pp. (1988 Reprint).

Wills, K., 2014: *Haunted Wiltshire.* The History Press. 96 pp.

Wilson, H. F., 1953: *The Jersey Shore: A Social and Economic History of the Counties of Atlantic, Cape May, Monmouth and Ocean.* 2 Volumes. New York: Lewes Historical Publishing. 1167 pp.

Witty, J. E., and Knox, E. G., 1989: "Identification, Role in Soil Taxonomy, and Worldwide Distribution of Fragipans." In Smeck, N. E., Ciolkosz, E. J. (eds.), *Fragipans: Their Occurrence, Classification, and Genesis.* SSSA Special Publication Number 24. Madison, WI: Soil Science Society of America. pp. 1–9.

Wolfe, L. J., Fox, S., Harris, R., Johnston, R, Jones, K., Manley, D., Tranos, E., and Wang, W. W., 2020: "Quantitative Geography III: Future Challenges and Challenging Futures." *Progress in Human Geography.* XX: X, 1–13. DOI: 10.1177/0309132520924722.

Wolfe, P. E., 1941: *Subsequent Topography, an Expression of Structure, Lithology and Soil.* Ph.D. dissertation, Princeton: Princeton University. 68 pp.

Wolfe, P. E., 1943: "Soil and Subsequent Topography." *Journal of Geology.* **51**: 204–211.

Wolfe, P. E., 1952: "Periglacial Frost-Thaw Basins in New Jersey (USA)." abstract and discussion by Tj. van Andel. In International Geological Congress, Congrès Géologique International, XIXe session- livrets guides des excursions au Maroc. Algiers, Algeria: f.15.

Wolfe, P. E., 1953: "Periglacial Frost-Thaw Basins in New Jersey." *Journal of Geology.* **61**: 133–141.

Wolfe, P. E., 1954: "Periglacial Frost-Thaw Basins in New Jersey (U.S.A.)." Congrès Géologique International, Comptes Rendus de la dix-neuvième session, Alger 1952, Section XIII, Questions diverses de générale, troisième partie paléontologie stratigraphique, Quaternaire et Pétrographie, Fascicule XV. p. 355.

Wolfe, P. E., 1956: "Pleistocene Periglacial Frost-Thaw Phenomena on the New Jersey Coastal Plain." *Transactions, New York Academy of Sciences.* **18**:

507–515.

Wolfe, P. E., 1977: *The Geology and Landscapes of New Jersey*. New York: Crane Russak. 351 pp.

Wolfe, S. A., Morse, P. D., Neudorf, C. M., Kokrlj, S. V., Lian, O. B., and O'Neill, H. B., 2018: "Contemporary Sand Wedge Development in Seasonally Frozen Ground and Paleoenvironmental Implications." *Geomorphology*. **308**: 215–229.

Wolfe, S. A., Murton, J., Bateman, M., and Barlow, J., 2020: "Oriented-Lake Development in the Context of Late Quaternary Landscape Evolution, McKinley Bay Coastal Plain, Western Arctic Canada." *Quaternary Science Reviews*. **242**: 106414.

Wolfe, S. A., Demitroff, M., Neudorf, C. M., Woronko, B., Chmielowska-Michalak, D., Lian, O. B., 2023: "Late Quaternary Eolian Dune-Field Mobilization and Stabilization near the Laurentide Ice Sheet limit, New Jersey Pine Barrens, Eastern USA." *Aeolian Research*. 62, June 2023, 100877.

Woo, Ming-Ko, 1993: "Northern Hydrology." In French, H. M., Slaymaker, O. (eds.), *Canada's Cold Environments*. Montreal and Kingston, Canada: McGill Queen's University Press. pp. 117–142.

Woo, Ming-Ko, 2012: *Permafrost Hydrology*. Berlin, Heidelberg: Springer, 564 pp.

Woo, Ming-Ko, and Pomeroy, J., 2012: "Snow and Runoff: Processes, Sensitivity and Vulnerability." In French, H. M., Slaymaker, O. (eds.), *Changing Cold Environments: A Canadian Perspective*. pp. 105–125.

Woo, Ming-Ko, and Xia, Z.. 1996: "Effects of Hydrology on the Thermal Conditions of the Active Layer." *Nordic Hydrology*. **27**: 129–142.

Woo, Ming-Ko, and Young, K. L., 2003: "Hydrogeomorphology of Patchy Wetlands in the High Arctic, Polar Desert Environment." *Wetlands*. **23**, 2: 291–309.

Woo, Ming-Ko, and Young, K. L., 2006: "High Arctic Wetlands: Their Occurrence, Hydrological Characteristics and Sustainability." *Journal of Hydrology*. **320**: 432–450.

Wood, P. J., Greenwood, M. T., and Agnew, M. D., 2003: "Pond Biodiversity and Habitat Loss in the UK." *Area*. **35**, 2: 206–216.

Woodbridge, S. W., and Morgan, R. S., 1937: *The Physical Basis of Geography: An Outline of Geomorphology*. London: Longmans, Green & Co. 445 pp.

Woodward, H. E., 1964: *Place Names of Salem County, New Jersey*. Salem County Historical Society Publications. **2**, 4: 1–85.

Woronko, B. J., and Bujak, L., 2018: "Quaternary Aeolian Activity of Eastern Europe (a Poland case study)." *Quaternary International*. **478**: 75–96.

Woronko, B., Karasiewicz, T. M., Rychel, J., Kupryjanowicz, M., Fiłoc, M., Moska, P., Adamczyk, A., and Demitroff, M. N., 2022a: "A Palaeoenvironmental Record of MIS 3 Climate Change in NE Poland—Sedimentary and Geochemical Evidence." *Quaternary International*. **617**: 80–100.

Woronko, B. J., Rychel, T. M., Karasiewicz, M., Kupryjanowicz, A., Adamczyk, M., Fiłoc, L., Marks, T., Krzywicki, K., and Pochocka-Szwarc, J., 2018: "Post-Saalian Transformation of Dry Valleys in Eastern Europe: An Example from NE Poland." *Quaternary International*. **467**: 161–177.

Woronko, B., Zagórski Z., and Cyglicki M., 2022b: "Soil-Development Differentiation Across a Glacial-Interglacial Cycle, Saalian Upland, E Poland." *Catena*. **211**, April, 105968.

Woronko, B., Zieliński, P., and Sokołowski, R. J., 2015: "Climate Evolution During the Pleniglacial and Late Glacial as Recorded in Quartz

Grain Morphoscopy of Fluvial to Aeolian Successions of the European Sand Belt." *Geologos.* 21: 89–103.

Worsley, P., 1985: "Periglacial Environment." *Progress in Physical Geography.* **10**: 265–274.

Worsley P., 1997: 2. "The Preservation Potential of Icings." *Quaternary Newsletter.* **83**: 52–54.

Worsley, P., 2016: "Sand Wedges in England and Arctic Canada." *Mercian Geologist.* **19**, 1: 7–17.

Wrangel, C. M., 1969: "Pastor Wrangel's Trip to the Shore. Translated and Edited by Carl Magnus Anderson from Wrangel's Day Book." *New Jersey History.* **87**: 5–31.

Wrench, K., (ed.), 2014: *Tar Heels: North Carolina's Forgotton Economy: Pitch, Tar, Turpentine & Longleaf Pines.* Createspace Independent Publishing Platform. 144 pp.

Wright, E., 1867a: Map of lands belonging to Stephan Colwell. Davis Collection # 229. 94.9.214.369. Hamilton Township Historical Society, Mays Landing, NJ, 1 sheet.

Wright, E., 1867b: Survey, A beginning corner Chas. Steelman to John Ford 1799 folio A in Book 372 Page 447 Mays Landing. Davis Collection #229, 94.9.229.384. Hamilton Township Historical Society, Mays Landing, NJ, 1 sheet.

Wright, E., c.1867: Survey, Mays Landing. Davis Collection. Hamilton Township Historical Society, Mays Landing, NJ, 1 sheet.

Wright, E., 1868: A survey of the Weymouth Tract showing exceptions, primarily from actual surveys made by E. Wright, surveyor and civil engineer. Traced from the original July 20th, 1891, Gloucester City, NJ Map #348, H28 Davis Site. 1 sheet.

Yershov, E. D., 1998: *General Geocryology.* Cambridge: Cambridge University Press. 580 pp.

Zampella, R. (A.). 2004: "The Kirkwood-Cohansey Project." *New Jersey Flows.* NJ Water Resources Research Institute Special Pinelands Edition. **5**, 2: 1.

Zampella, R. A., Bunnell, J. F., Laidig, K. J., and Procopio, N.A. III, 2010: "Aquatic Degradation in Shallow Coastal Plain Lakes: Gradients or Thresholds?" *Ecological Indicators.* **10**: 303–310.

Zampella, R. A., and Roman, C. T., 1983: "Wetlands Protection in the New Jersey Pinelands." *Wetlands.* **3**: 124–133.

Zanner, C. W., 1999: *Late-Quaternary Landscape Evolution in Southeastern Minnesota: Loess, Eolian Sand, and the Periglacial Environment.* PhD dissertation, University of Minnesota, Minneapolis, MN, 395 pp.

Zhang, A. M., Feng, Z., Wang, X., Sun, H., Zhao, J., and U, H., 2007a: "Sandy Grassland Blowouts in Hulunbuir, Northeast China: Geomorphology, Distribution, and Causes." *Progress in Natural Science.* **17**, 1: 68–73.

Zhang, A. M., Wang, X., U, H., and Feng, Z., 2007b: "HulunBuir Sandy Grassland Blowouts (III): Influence of Soil Layer and Microrelief." *Journal of Desert Research.* **27**, 1: 26–31 (in Chinese with English abstract).

Zielinski, T., and Gozdzik, J., 2001: "Palaeoenvironmental Interpretation of a Pleistocene Alluvial Succession in Central Poland: Sedimentary Facies Analysis as a Tool for Palaeoclimatic Influences." *Boreas.* **30**: 240–253.

Zinkin, V., 1976: *Place Names of Ocean County, New Jersey: 1609–1849.* Toms River, NJ: Ocean County Historical Society. 214 pp.

Colophon

Abriana Bradley and Tom Kinsella completed the editing, design, and typesetting for *Soggy Ground*. Body text is 12-point Lora; chapter titles are 20-point Montserrat; footnotes employ 10-point Lora. Gary Schenck completed the cover artwork.

This is a publication of the South Jersey Culture & History Center at Stockton University. Our mission is to foster awareness within local communities of the rich cultural and historical heritage of southern New Jersey, to promote the study of this heritage, especially among area students, and to produce publishable materials that provide a lasting and deepened understanding of this heritage.

stockton.edu/sjchc/